MEDICINAL PLANTS IN ANDHRA PRADESH INDIA

T. PULLAIAH
Department of Botany
Sri Krishnadevaraya University
Anantapur – 515003
Andhra Pradesh
India

Regency Publications
NEW DELHI

ISBN 81-87498-51-X

Published by Regency Publications, 20/36-G, Old Market, West Patel Nagar, New Delhi 110 008 and printed at Radiant Printers, New Delhi, Phones: 574 0038, 571 2539, 406 0485; Telefax: 578 3571.

Email: regency@satyam.net.in

PREFACE

Andhra pradesh is one of the 28 states of the Indian union. Situated in the middle portion of Eastern half of the Indian peninsula, the state lies between the latitudes 12°37′ N and 19°54′ N and longitudes 76°46′ E and 84°46′ E. The total area of the state is 2,75,068 sq.km. It has 968 km sea coast line along the Bay of Bengal. The altitude ranges from sea coast to 1500 m above MSL. Annual rainfall ranges between 54 cm in south west to 200 cm in north east. Maximum temperature in summer ranges between 37°C and 44°C while the minimum temperature in the winter ranges between 14°C and 19°C. 412 medicinal plants have been enumerated.

The following sequence has been followed for the medicinal plants:

Valid name followed by Synonym, Family, brief description, distribution in Andhra Pradesh, flowering and fruiting season and medicinal uses. Active principle or chemistry has been given to some plants.

I thank my wife T. Amrutavalli for her help in preparation of the manuscript.

Abelmoschus manihot (L.) Medikus

Syn: *Hibiscus manihot* L.
Hibiscus pungens Roxb.

Family: Malvaceae.

Perennial, simple or sparingly branched, strigose and weakly prickly undershrubs, up to 3m tall. Leaves broadly ovate. Flowers yellow with a deep purple or chocolate base within, in terminal racemes. Fruit capsule, ovoid-oblong.

Rare in north coastal Andhra Pradesh.

Fl.& Fr.: Aug–Jan.

Propagation: by seeds.

Medicinal uses: Roots are used for tuberculosis and the mucilage has hypoglycaemic activity. The viscous substance Abelmosan A,B and C are isolated from the roots. These substances are useful as antidiabetic and for prevention of liver failure.

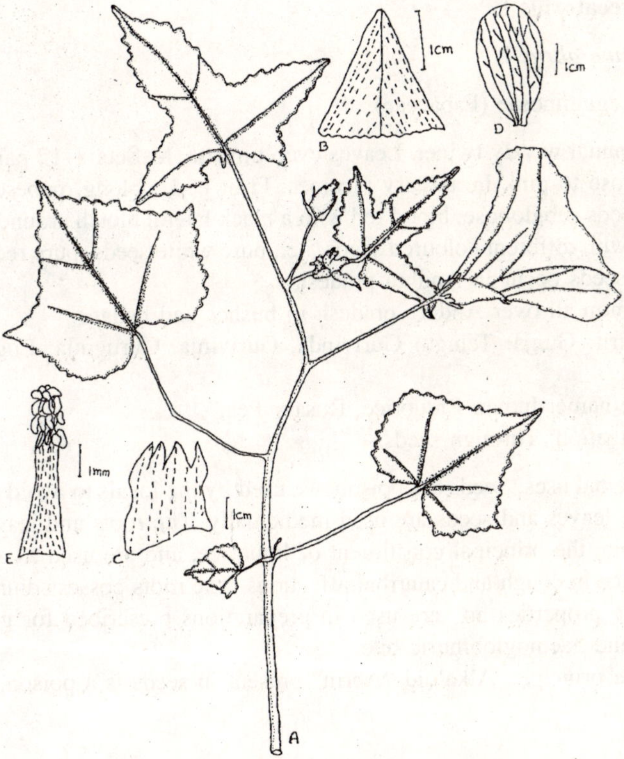

Fig. 1: *Abelmoschus moschatus* Medikus
A. Twig; B. Calyx lobe; C. Calyx split open; D. Petal; E. Staminal column

Abelmoschus moschatus Medikus

Syn: *Hibiscus abelmoschus* L.

Family: Malvaceae

An annual, erect herb, up to 1.5m tall, branchlets hairy. Leaves 3–5-lobed, pubescent. Flowers bright yellow with dark purple base. Fruit capsule, fusiform, hairy, seeds many, black-greyish brown, concentrically ribbed, scented.

Occasional in Kurnool, Prakasam and Khammam districts.

Fl.& Fr.: July–Dec.

Sanskrit: Kasturilatika; Telugu; Kasturi Benda.

Trade Name: Ambrette plant, Muskmullon

Propagation: by seeds.

Medicinal uses: The seeds are useful in perfumary and medicine. The seeds are used as a tonic, stimulant, carminative. The crushed seeds, on steam distillation yield a volatile oil, known as musk seed oil or Amberette seed oil.

Abrus precatorius L.

Syn: *Glycine abrus* L.

Family: Leguminosae (Fabaceae)

An annual woody twiner. Leaves even pinnate, leaflets 7–12 pairs, oblong. Flowers rose to pink in axillary racemes, Fruit pod, oblong, pubescent, 5–10-seeded, seeds subglobose; blood red with a black lateral blotch around the hilum. Varieties with different coloured seeds, i.e., pure white seeds, pure red seeds and all black seeds occur in Andhra pradesh.

Common all over Andhra pradesh in bushes and hedges.

Sanskrit: Gunja; Telugu: Gurivinda, Gurivinja, Guruginja; English: Wild Liquorice.

Trade name: Indian Liquorice, Rosary Pea.

Propagation: Through seeds.

Medicinal uses: Seeds are poisonous, used by the locals to avoid pregnancy. The roots, leaves and seeds are used medicinally. The roots and leaves contain glycyrrhizin, the principal constituent of liquorice, and are used as a substitute for liquorice in cough and catarrhal affectious. The roots possess diuretic, tonic, and emetie properties an are used in preparations prescribed for gonorrhoea, jaundice and haemoglobinuric bile.

Active principle: Alkaloid "Abrin" present in seeds is a poison.

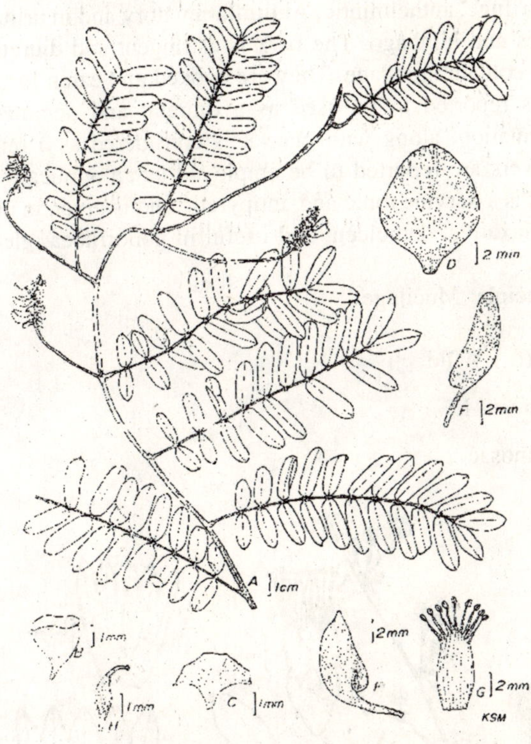

Fig. 2: *Abrus precatorius* L.
A. Twig; B. Calyx; C. Calyx split open; D. Standard petal; E. Wing petal; F. Keel petal; G. Staminal column; H. Pistil.

Abutilon indicum (L.) Sweet

Syn: *Sida indica* L.

Family: Malvaceae

Perennial woody densely tomentose undershrub, up to 1m tall. Leaves broadly ovate-orbicular, deeply cordate, pubescent. Flowers orange yellow, axillary, solitary. Fruit schizocarp, depressed-globose, tomentose. Seeds black, tubercled.

Common in hedges and wastelands all over Andhra Pradesh.

Sanskrit: Atibala. Telugu: Thuthurubenda, Papidlakaya, Kammalaku.

Trade name: Countr mallow.

Propagation: Through seeds.

Nervine tonic. Seeds aphrodisiac. Used in Dysentry, Leucorrhoea, Piles, Psoriasis, Diuretic and as Galactogogue

Medicinal uses: Country mallow is used as drug in both Ayurvedic and Unani Systems. The drug is available as dried whole plant. This is reported to be useful as febrifuge, anthelmintic, Antiinflammatory and in urinary and uterine discharges, piles and lumbago. The bark is astringent and diuretic.

The leaves contain mucilage. They are cooked and eaten for bleeding piles. Their extract is reported to be used as a diuretic and demulcent and as an emollient fomentation; along with ghee it is considered as a remedy for diarrhoea. The flowers are reported to be employed as an application to boils and ulcers. The root is a nervine tonic and antipyretic and is given in piles. The seeds are considered laxative, demulcent and useful in gonorrhoea, gleet and chronic cystitis.

Active principle: Mucilage.

Acacia caesia (L.) Willd.

Syn: *Mimosa caesia* L.

Family: Leguminosae

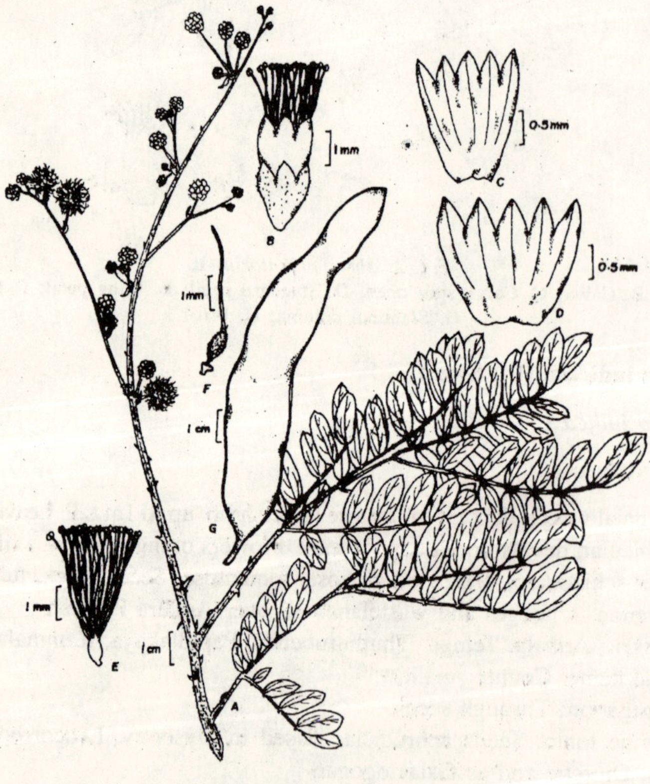

Fig . 3: *Acacia caesia* (L.) Willd.
A. Twig ; B. Flower; C. Calyx split open; D. Corolla split open; E. Staminal column; F. Pistil; G. Pod.

Armed woody straggling shrub, up to 6m tall, internodal thorns hooked; Leaves bipinnate, pinnae 6–8 pairs, leaflets 15–25 pairs, oblong, pubescent. Flowers in globose heads, creamish yellow. Fruit pod, oblong, glabrous, horned, 10–12-seeded. Occasional in dry deciduous forests of all districts.

Sanskrit: Nikunjika; Telugu: Yerra cheeki.

Trade name: Soap bark.

Propagation: by seeds.

Medicinal uses: Bark has cleaning properties and protects the skin against microorganisms. Flowers used by women to treat menstrual disorders.

Acacia catechu (L.f.) Willd.

Syn: *Mimosa catechu* L.f.

Family: Leguminosae (Fabaceae).

Shrub like armed trees, 4–5m tall, young branchlets dark brown to purple, 2 curved stipular spines. Leaves 2-pinnate, pinnae 10–15 pairs, leaflets 20–40 pairs, elliptic-oblong. Flowers light greenish-yellow in axillary spikes. Pod flat, thin, straight, glabrous, dark-brown, shining, seeds 4–10, greyish brown, shining. Occasional in dry deciduous forests.

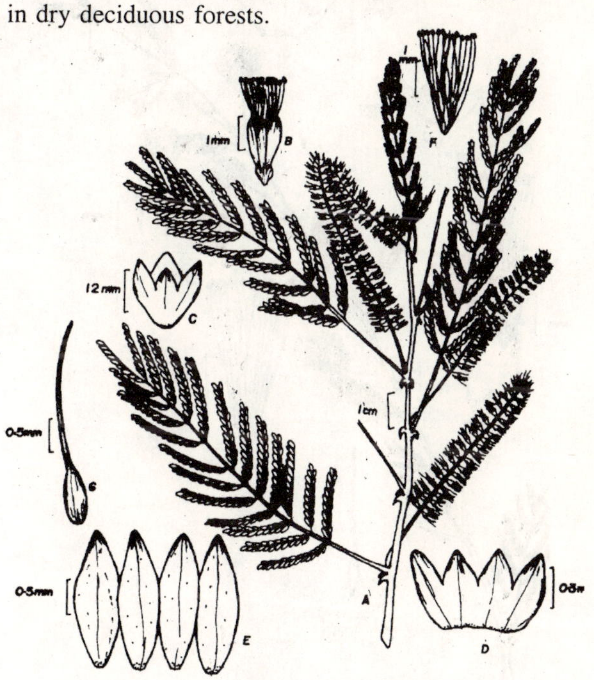

Fig. 4: *Acacia catechu* (L.f.) Willd.
A. Twig; B. Flower; C. Calyx entire; D. Calyx split open; E. Corolla split open; F. Staminal column; G. Pistil.

Sanskrit: Khadira. Telugu.: Nalla sandra. English; Catechu.
Trade name: Cutch tree.
Fl.& Fr.: August–November

Medicinal uses: Bark and heart wood used in conjunctivitis, Haemoptysis, Appetiser, Leprosy, Leucoderma, Haemorrhage, Diabetes, Anthelmintic, Tonic.

Leaves are used for dysentery, flowers for gonorrhoea. Bark is useful as astringent and abortifacient. Decoction of the bark is useful for scabies and other chronic skin troubles.

Active principle: Catechin, Catechutannic acid and Tannin. It is given in child birth, diarrhoea, haemoptysis and leucorrhoea. The plant is used for ashma, bronchitis, pain in chest, sore in mouth, colic and cancer. Heartwood is inhaled to check bleeding from nose. Root is used in rheumatism and toothache.

Acacia nilotica (L.) Willd. ex Del.

Syn: *Acacia arabica* auct non (Lam.) Willd.
Mimosa arabica Lam.

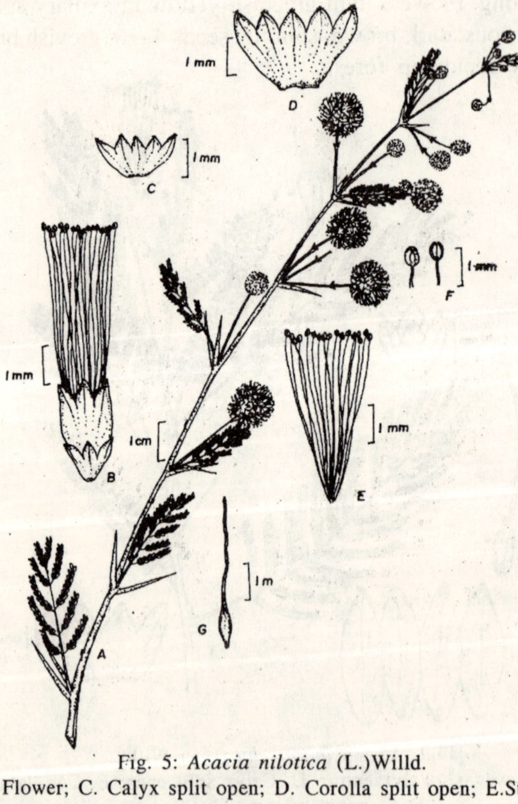

Fig. 5: *Acacia nilotica* (L.)Willd.
A. Twig; B. Flower; C. Calyx split open; D. Corolla split open; E.Staminal column; F. Anthers; G. Pistil.

Family: Leguminosae (Fabaceae).

Armed tree, up to 15m tall, bark dark brown or black, longitudinally fissured or deeply cracked. Leaves bipinnate, pinnae 4 pairs, leaflets 15–20 pairs, oblong. Flowers golden yellow in axillary globose heads. Fruit pod, moniliform, whitish tomentose, constricted between seeds, seeds 8–12, dark brown, compressed, shining.

Common all over Andhra Pradesh, in waste lands, hedges and tank bunds.
Fl.: June–September, Fr.: ripen in April–May.
Sanskrit: Barburah, Telugu: Nallatumma. Eng: Indian Gum Arabic.
Trade name: Babul.
Propagation: By seeds.

Medicinal uses: Bark is used for asthma, bronchitis, diabetes, dysentery, diarrahoea and skin diseases. Stem is used for tooth brush and gum for burns. Leaves are useful for eye complaints, headache, throat infection, urinary problems and gonorrhoea. The flowers are used for ear complaints and as tonic. Bark is used in haemorhages, wounds, ulcers, also used as a decoction in the treatment of diarrhoea and vaginal secretions.
Aphrodisiac, Diuretic, used in Leprosy, Leucoderma, Bronchitis, Seminal weakness, Utero-Vesical disorders.
Active principle: Tannin and phenolic compounds.

Acacia sinuata (Lour.) Merr.

Syn: *Acacia concinna* (Willd.) DC.
Mimosa sinuata Lour.
Mimosa concinna Willd.

Family: Leguminosae (Fabaceae)

Prickly climbing shrubs, internodal thorns hooked. Leaves bipinnate, pinnae 8–15 pairs, leaflets 20–25 pairs, linear, pubescent. Flowers creamy-white in terminal heads. Fruit pod, fleshy, wrinkled when dry, depressed between seeds, seeds 10–14.

Common in forests of most districts.
Fl.: & Fr.: Sep–March.
Propagation: By seeds.
Sanskrit: Saptala, Amala, Charmakasha. Telugu;.: Shikaya. English: Soap pod.

Medicinal uses: Pods Diuretic, Anthelmintic, Haemorrhoids, Leucoderma, Leprosy, Abscesses, Eczema, skin diseases, Bone fracture.

An infusion of the leaves is used in malarial fever. A decoction of the pods relieves biliousness and acts as a purgative. It is used to remove dandruff. An ointment prepared from the ground pods is good for skin diseases.

Active principle: Saponin and Alkaloid.

Acalypha indica L.

Family: Euphorbiaceae.

An annual erect herb, up to 75cm tall. Leaves ovate-rhomboid. Flowers unisexual, green, in axillary spikes, male in upper half, female in the lower half. Fruit capsule, globose.

Common weed in waste lands all over Andhra Pradesh.

Fl.& Fr.: All round the year.

Sanskrit: Haritamanjari. Telugu: Muripindi, Kuppintaku.

Trade name: Indian Acalypha.

Propagation: by seeds.

Medicinal uses: The herb is said to possess diuretic, carminative, expectorant and emetic properties, but it causes gastro intestinal irritation. It is used as a safe and speedy laxative and also to cure tooth and earache. In Homeopathy the herb is used as a remedy for severe cough associated with bleeding from the lungs, haemoptysis and incipient phthisis.

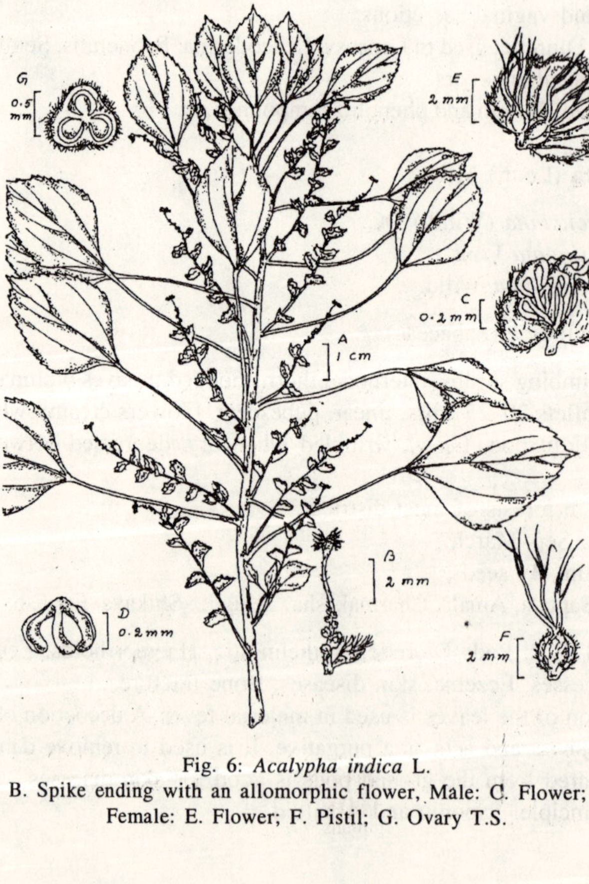

Fig. 6: *Acalypha indica* L.
A. Twig; B. Spike ending with an allomorphic flower, Male: C. Flower; D. Stamen; Female: E. Flower; F. Pistil; G. Ovary T.S.

A paste of leaves is applied to burns; with lime juice it is useful in early cases of ring worm. Fresh juice of the leaves is applied with oil, salt or lime in rheumatoid arthritis and to cure scabies and other skin affections. The powdered leaves are used for bedsores and maggot-infested wounds.

Active principle: Cyanogenic glucoside and alkaloids. viz: acalyphine and triacetoneamine.

Acanthus ilicifolius L.

Family: Acanthaceae

A gregarious shrub reaching up to 1.6m tall with stilt roots. Leaves oblong or elliptic, pinnatifid, lobes zig zag, spinous. Flowers large and blue in long slightly interrupted spikes. Fruit ellipsoid capsule.

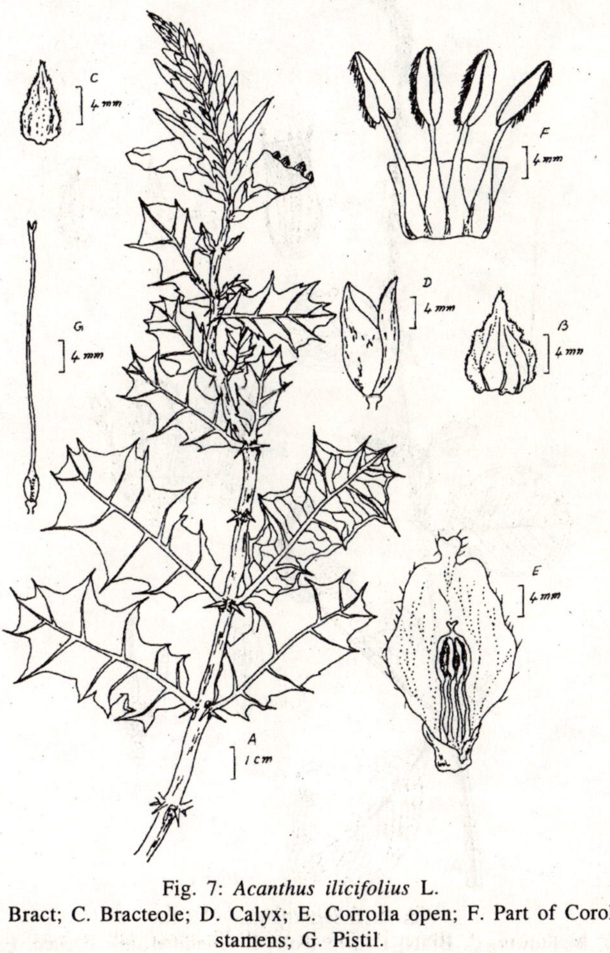

Fig. 7: *Acanthus ilicifolius* L.
A. Twig; B. Bract; C. Bracteole; D. Calyx; E. Corrolla open; F. Part of Corolla showing stamens; G. Pistil.

Common in tidal forests of coastal districts.
Fl.: Fr.: April–July.
Sanskrit: Harikrishna. Telugu: Alchi.
Trade name: Sea holly.
Propagation: by seeds.

Medicinal uses: A decoction of the plant with sugar candy and cumin is used in dyspepsia with acid. It is also considered to be a diuretic and is used as a cure for dropsy and bilious swellings. Leaves are used as an expectorant.

Achyranthes aspera L.

Family: Amaranthaceae.

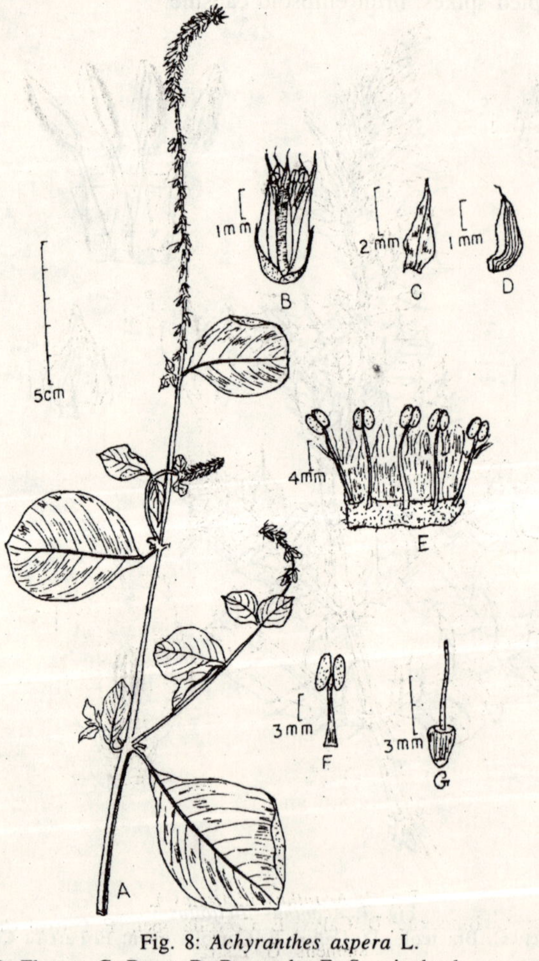

Fig. 8: *Achyranthes aspera* L.
A. Twig; B. Flower; C. Bract; D. Bracteole; E. Staminal tube opened; F. Stamen; G. Pistil.

Annual erect herb, up to 1m tall,. Leaves orbicular-obovate. Flowers greenish-white in terminal spikes. Fruit utricle, ovoid.

Common weed of wastelands and hedges.

Fl.: Fr.: All round the year.

Sanskrit: Apamarga; Telugu: Uthareni

Propagation: by seeds.

Trade name: Prickly-chaff Flower.

Medicinal uses: The plant is reported to be pungent, astringent, pectoral and diuretic. It is used as an emmengogue and in piles and skin eruption. A decoction of the plant is useful in pneumonia and renal dropsy; in large doses, however the decoction or juice acts an ecbolic.The juice of the plant is reported to be used in opthalmia and dysentery.

Active principle: Alkaloid achyranthine and betaine.

Achyranthes bidentata Bl.

Family: Amaranthaceae.

Annual erect herb, up to 1.5m tall covered with fine hairs. Leaves elliptic-oblong or ovate-lanceolate, sparsely hairy above, densely hairy beneath. Flowers in terminal elongated spikes. Fruit utricle, seeds black.

Rare in north coastal districts.

Fl. & Fr.: July–November.

Sanskrit: Apamarga.

Propagation: by seeds.

Medicinal uses: Leaf is used for treating blisters in the mouth and also for cholera. Root is used to treat scorpion stings. Whole plant is used against whooping cough. The plant is diuretic and astringent.

Acorus calamus L.

Family: Araceae

Semi aquatic herb with a creeping and branched aromatic rhizome. Leaves bright green, arranged alternately in two vertical rows on the rhizome. Flowers light brown in sessile cylindrical spadix. Fruit oblong berry.

It is found wild or cultivated throughout India.

Sanskrit: Bhuta nashini. Telugu: Vasa. English: The sweet Flag.

Native of south Asia and Central and Western North America.

Medicinal uses: The dry rhizome contains some yellow aromatic oil. The dried rhizome is known to have emetic, carminative, anthelmintic and antispasmodic properties and also gives good results in dyspepsia and chronic diarrhoea and dysentry. Root extract exhibts antimicrobial activity against *Staphylococcus*, *E.coli* and *Asperigillus niger.*

Active principe: Glucoside Acorin and Essential oil. Tranquilizer.

Note: Item banned for export.

Actinodaphne madraspatana Bedd. ex Hook.f.

Family: Lauraceae.

Medium sized tree, branchlets pubescent. Leaves elliptic-lanceolate, coriaceous, shining above. Flowers dioecious, yellowish from the naked branches, male flowers in clusters, female flowers umbellate. Fruit ellipsoid.

Occasional in forests in Rayalaseema and north coastal districts.

Fl. & fr.: February–April.

Telugu: Kovangutti.

Trade name: Ray Laurel.

Propagation: by seeds.

Medicinal uses: Leaves, flowers and fruits constitute the drug. It is precursor of Vitamin A. It is reported to be used to cure mania and fickle minded behaviour and to clear wounds.

Actinopteris radiata (Swartz.) Link

Syn: *Actinopteris australis* (L.f.) Link

Family: Actinopteridaceae

Terrestrial fern. Rhizome very short. Lamina fan-shaped, all arising from the rhizome, pinnae clustered, deeply dissected, lobes linear, glabrous, Sori arranged in two rows on the lower side of the pinnae lobes.

Common all over the state under rocks in forests.

Trade name: Peacock's tail.

Medicinal uses: The plant is used in bronchitis and gynaecological disorders. The plant is administered in burns, wounds and scalds. The dry leaves are used in tuberculosis.

Adhatoda zeylanica Medik.

Syn: *Adhatoda vasica* Nees
Justicia adhatoda L.

Family: Acanthaceae

Erect shrub, up to 1.5 m tall, branchlets pubescent. Leaves oblanceolate. Flowers creamish white in dense, axillary spikes. Fruit capsule, beaked, seeds 4, orbicular, rugose.

Common in all plains districts near villages.

PLATES

Tacca leontopetaloides (L.) O.Kuntze

Tinospora cordifolia (Willd.) Miers ex Hook. f.

Tectona grandis Lf.

Zaleya decadra (L.) Burm.f.

Wrightia tinctoria R.Br.

PLATE XIV

Fl.& Fr.: November–March.
Sanskrit: Vasa,Vasaka. Telugu: Addasaramu.
Trade name: Vasaka.

Medicinal uses: The leaves, flowers, fruits and roots are extensively used for treating cold, cough, whooping cough and chronic bronchitis and asthma, as sedative-expectorant, antispasmodic and as anthelmintic. In chronic bronchitis, it is efficacious and affords relief, especially when sputum is thick and tenacious; The cough is relieved and sputum is liquified and is easily expelled.

Active principle: Leaves contain alkaloids Vasicine and vasicinone. The leaf-juice is stated to cure diarrhoea, dysentery and glandular tumor, and is given as emmengogue.

Adiantum capillus-veneris L.

Family: Adiantaceae.

Rhizomatous erect herbaceous fern, up to 30cm tall, rhizome densely scaly. Fronds (leaves)2–3-pinnate, the lowest pinna slightly branched again, pinnula glabrous, outer margins incised. Sori elliptic or linear, arranged in roundish sinuses one to each lobe.

Occasional along the streams in most districts.
Trade name: The Maidenhair Fern.

Medicinal uses: The herb is used as a stimulating pectoral, emmengogue, purgative, demulcent, emollient and tonic and also in catarrhal affections.

Aegle marmelos(L.) Correa

Syn: *Crateva marmelos L.*

Family: Rutaceae

Moderate sized, slender, aromatic, armed tree, 6 to 10m tall. Bark grey, exfoliating with longitudinal irregular flakes. Leaves 3(5)-foliolate, leaflets ovate-elliptic. Flowers greenish-white in axillary panicles, sweet scented. Fruit ovoid-globose, greenish white, rind woody, seeds-numerous embedded in aromatic pulp.

Common in deciduous forests all over Andhra pradesh.
Fl. & Fr.: March–May. Ripen the following year.
Sanskrit: Bilva, Sriphal; Telugu: Maredu, Bilva
Trade name: Holy fruit tree.

Medicinal uses: The unripe or half ripe fruit is regarded as astringent, digestive and stomachic. The fruit is used in chronic diarrhoea and dysentery and said to act as a tonic for heart and brain. Clinical trials of unripe fruits showed anti-viral activity against Ranikhet disease virus, hypoglycaemic activity and significant results against intestinal parasities. The pulp diluted with water and added

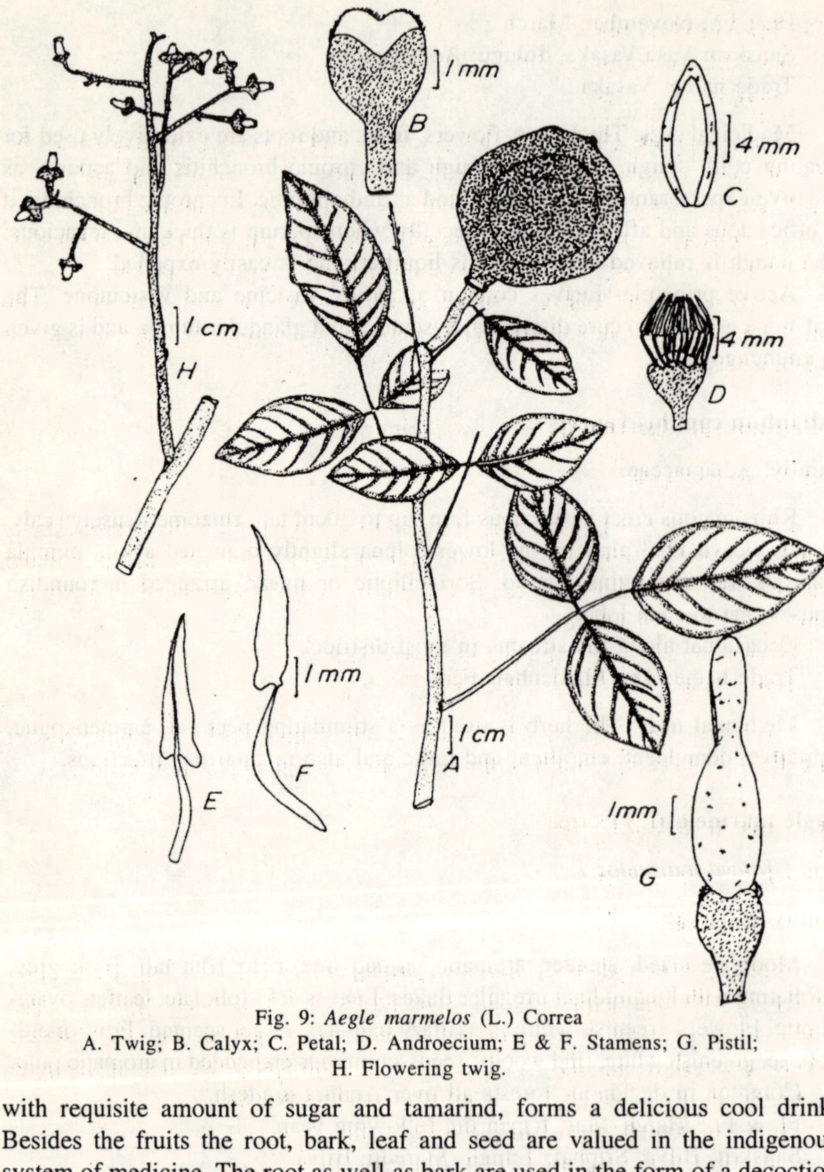

Fig. 9: *Aegle marmelos* (L.) Correa
A. Twig; B. Calyx; C. Petal; D. Androecium; E & F. Stamens; G. Pistil; H. Flowering twig.

with requisite amount of sugar and tamarind, forms a delicious cool drink. Besides the fruits the root, bark, leaf and seed are valued in the indigenous system of medicine. The root as well as bark are used in the form of a decoction as a remedy in melancholia, intermittent fevers and palpitation of the heart. Poultice made of leaves are used for opthalmia and ulcers. Leaves are used for abortion and steility.

Active principle: Tannins, Saccharides, Furocoumarins, Furowuinolin alkaloids.

Aerva lanata (L.) A.L.Juss.

Syn: *Achyranthes lanata* L.

Family: Amaranthaceae.

Annual erect or suberect herb, up to 75cm tall. Leaves alternate to subopposite, ovate, pubescent above, white woolly beneath. Flowers greenish-white in dense axillary spikes. Seeds annular, testa shiny, black.

Occasional weed in cultivated fields, waste lands and hedges all over Andhra pradesh.

Fl.& Fr.: Through out the year.

Sanskrit: Pashanabheda, Bhadra, Telugu: Konda pindi, Pindikura

Propagation: By seeds.

Medicinal uses: The plant is anthelmintic and demulcent and is used in lithiases. It is also regarded as a valuable medicine for cough, sore throat, indigestion and wounds. A decoctin of the plant is an efficaceous diuretic and is useful in catarh of bladder. The flowers are used for removal of kidney stones and in gonorrhoea. The roots are diuretic and demulcent. The roots and flowers are used to cure headache.

Agave americana L.

Family: Agavaceae.

A perennial stemless, scapigerous shrub. Leaves very large, lanceolate. Inflorescence stock 4–8m long, apically repeateldly branched. Flowers densely crowded with strong evil smell.

Common in Rayalaseema in hedges and degraded scrub jungles.

Fl. & Fr.: September–January

Sanskrit: Kalakantala. Telugu: Kalabanda, Kittanara.

Trade name: Century plant, American Aloe.

Propagation: By suckers.

Medicinal uses: The leaves are used as laxative and emmenagogue and in scurvy. They are also used as a resolvent in syphilis, scrofula and cancer. Thin slices of leaves are employed as poultice. The roots possess diuretic, diaphoretic and antisyphilitic properties.

Active principle: Leaves contain Hecogenin, a sterol.

Ageratum conyzoides L.

Family: Asteraceae (Compositae).

Annual erect pubescent herbs, up to 75cm tall. Leaves ovate. Florets pale blue in terminal homogamous heads; Fruit achene, black.

A common weed of irrigated fields, streams and moist localities in all districts.

Fl. & Fr.: July–January.
Telugu: Pumullu.
Trade name: Goat weed, White weed.
Propagation: by seeds.

Medicinal uses: A decoction infusion of the herb is given in stomach ailments such as diarrhoea, dysentery and intestinal colic with flatulence and also in rheumatism and fever. The leaves are used for fomentation in leprosy and other cutaneous diseases. The juice of the leaves and roots is reported to be used as an eye lotion.

Ailanthus excelsa Roxb.

Family: Simaroubaceae

Large trees, up to 25m tall with rough bark and soft white wood, branchlets with persistent leaf-scars. Leaves imparipinnate, leaflets 8–14 pairs, subopposite, oblong-lanceolate, inequilateral at base. Flowers large, polygamous, in axillary and terminal panicles. Fruit samara, oblong, coppery red.
Occasional in forests.
Fl.: January–March. Fr.: March–May.
Sanskrit: Madala, Mahaniba. Telugu: Peddamanu, Peddavepamanu.
Trade name: Tree of Heaven.
Propagation: by root cuttings and seeds.

Medicinal uses: The bark is bitter, astringent, febrifuge and anthelmintic. It has antispasmodic and expectorant properties and is used for asthma, bronchitis and dysentery. It is used for dyspepsia and earache and has antiseptic properties.

Alangium salvifolium(l.f.) Wang.

Syn: *Alangium lamarckii* Thw.
Grewia salvifolia L.f.

Family: Alangiaceae

Deciduous stragglers, shrubs or trees, up to 20m tall, branchlets at times spinescent. Leaves ovate-oblong to elliptic-lanceolate, chartaceous. Flowers white-cream, fragrant, densely rusty tomentose, often in clusters. Fruit berry, red when ripe.
Common in most districts in forests.
Trade name: Sage leaved alangium.
Propagation: by seeds.

Medicinal uses: The roots have been in use in Ayurvedic medicine. The root bark is very bitter and is reputed as a cure for skin diseases. It is credited with astringent, pungent, anthelmintic, purgative and emetic properties and prescribed

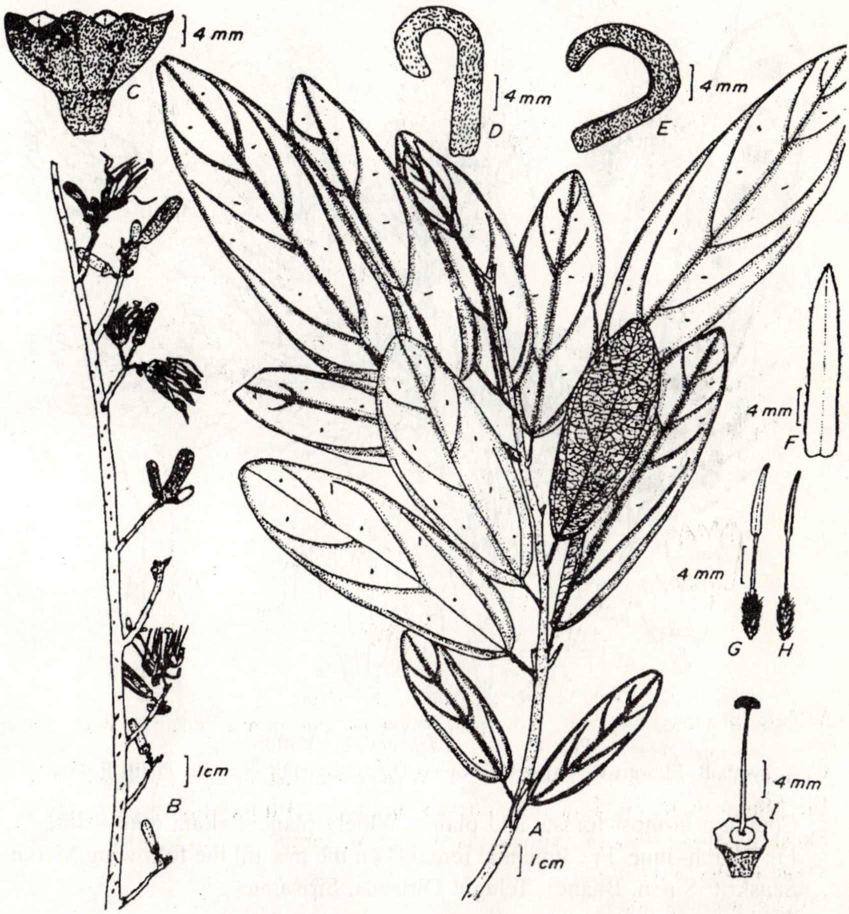

Fig. 10: *Alangiam salvifolium* (L.f) Wang.
A. Twig; B. Inflorescence branch; C. Calyx; D., E. & F. Petals; G. & H. Stamens; I. Pistil.

for biliousness and colic. The leaves are applied as poultice and rheumatism.

Active principle: The root bark contains alkaloids alangine A &B, alangicine, Cephealine, emetine, lamarckimine etc.

Albizia lebbeck (L.) Benth.

Syn: *Mimosa lebbeck* L.

Family: Leguminosae-Mimosaceae

Large deciduous tree, bark brownish-grey. Leaves bipinnate, pinnae 2–4 pairs, leaflets 4–8 pairs, oblong-ovate. Flowers greenish yellow in heads, fragrant. Fruit pod, pale yellow, linear-oblong, strap-shaped. Seeds 2–12, pale brown, flattened.

Fig. 11: *Albizia lebbeck* (L.) Willd.

A. Twig; B. Flower; C. Calyx split open; D. corolla split open; E. Pistil; F. Pod.

Common in most forests and plains. Widely planted along road sides.

Fl.: March–June. Fr.: June and remains on the tree till the following March.

Sanskrit: Sirish, Bhandi. Telugu: Dirisana, Sirisamu.

Trade name: Siris tree.

Propagation: by seeds.

Medicinal uses: The plant is reported to have antiseptic, antidysenteric and anti tubercular properties. The bark has acrid taste. It is recommended for bronchitis, leprosy, paralysis and helminth infections. The bark and seeds are astringent, useful in piles and diarrhoea and act as tonic. The root bark and root gum are used as dental powder for strengthening the gums.

Albizia odoratissima (L.f.) Benth.

Syn: *Mimosa odoratissima* L.

Family: Mimosaceae

Large deciduous tree, 15–25m tall. Bark dark grey. Leaves bipinnate, pinnae 2–5 pairs, leaflets 4–15 pairs, obliquely oblong. Flowers pale yellowish white,

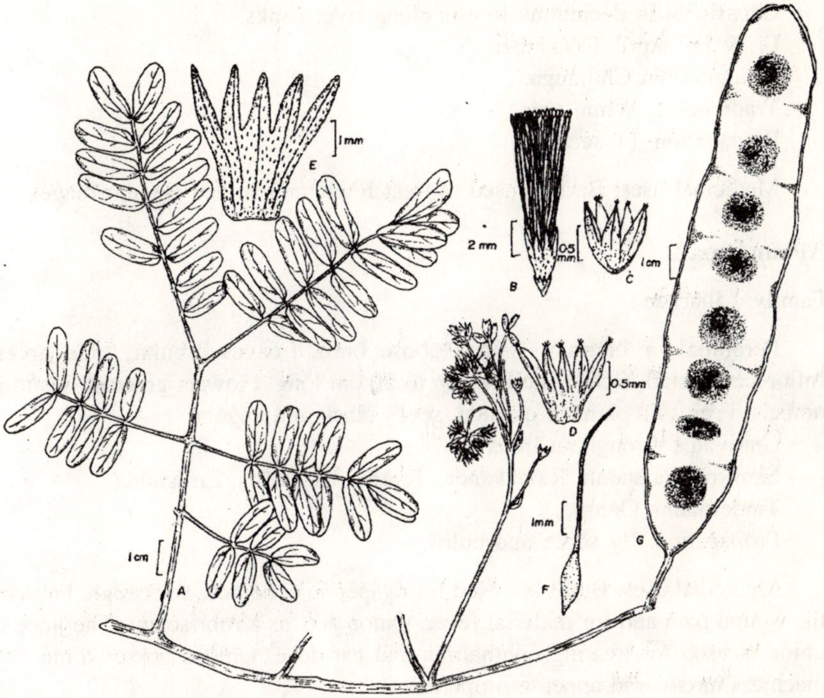

Fig. 12: *Albizia odoratissima* (L.f.) Benth.
A. Twig; B. Flower; C. Calyx entire; D. Corolla entire; E. Corolla split open; F. Pistil; G. Pod.

fragrant in terminal heads. Fruit oblong pod, reddish brown at maturity.

Common in deciduous forests.

Fl.: April–June; Fr: June–Dec.

Sanskrit:Bhusirisah. Telugu: Chinduga

Trade name: Black siris, kala siris.

Propagation: by seeds

Medicinal uses: The bark is used for curing insect bites. It is also used to cure ulcers, leprosy, skin diseases, cough, bronchitis, diabetis and burning sensation.

Albizia procera (Roxb.) Benth.

Syn: *Mimosa procera* Roxb.

Family: Mimosaceae.

Very large deciduous trees, up to 30m tall. Leaves pinnately compound, pinnae usually 3 pairs, leaflets 6–12 pairs, oblique, ovate. Flowers in globose heads. Fruit pod, orange-brown, seeds 10–12, yellow.

Occasional in deciduous forests along river banks.
Fl. & Fr.: April–December.
Telugu: Tella Chinduga.
Trade name: White siris.
Propagation: by seeds.

Medicinal uses: Bark is used to treat Rheumatism, and Haemorrhages.

Allium cepa L.

Family: Liliaceae

Perennial or biennial with globose bulb. Leaves tubular, blue green. Inflorecence stalk (scape) hollow, up to 50 cm long. Flowers greenish white in umbels. Fruit thin skinned capsule, seeds black and angular.
Cultivated throughout India.
Sanskrit: Palanduh, Raktakanda. Telugu: Ulligadda, Erragadda
Trade name: Onion.
Propagation: By seeds and bulbs.

Medicinal uses: Bulbs are used for dyspeptic condition, for cough, bronchitis, wound pain and for malarial fever. Onion acts as a rubifacient. The juice of onion is used for treating ophthalmia and ear ache. Onions posses tonic, stomachic, diuretic and appetite properties.

Active principle: Allins (alkylcysteine sulphoxides) in particular allylallin, Flavonoids. Onion is reported to posses aphrodisiac and antirheumatic properties. It is used for diseases of tumors and spleen. Onions are useful in jaundice and biliousness as they stimulate bile production. Onions are stimulant and expectorant and are, therefore, useful in respiratory affections.

Allium sativum L.

Family: Liliaceae

Perennial herb, 25–70 cm tall, stem erect, leafy in the middle. Leaves flat, 4–25cm long. Garlic bulb is usually a compound bulb, oval in shape, the skin colour of the bulb is silky white. Flowers reddish or greenish white, in cluster.
Cultivated all over the state.
Sanskrit: Arishtha, Shuklakanda. Telugu: Vellulli, Tellagadda.
Trade name: Garlic.
Propagation: by seeds.

Medicinal uses: Garlic is stimulant, diaphoretic, expectorant, diuretic and tonic. It is rubifacient when applied externally. It is also used as an anti-tubercular drug and in the treatment of epilepsy. It has also Anticancer activity. Bulbs are used in bronchitis, constipation, joint pain and fever. It is used in the form of juice, syrup or poultice. Garlic is used in the treatment of hypertension, diabetes, helminthiasis, chronic colitis, bacterial and fungal infections, amoebiosis and rheumatoid arthritis.

Active principle: Allins (Alkyl cysteine sulfoxide) especially allylallin, propenylallin and methylallin.

Aloe vera (L.) Burm. f.

Family: Liliaceae

A stoloniferous herb. Leaves dense, aggregated, succulent, ensiform, pale green with distant horny prickles on the margins. Scape 60–90cm long, simple. Raceme dense, flowers reddish-yellow, bisexual. Fruit capsule, ellipsoid.

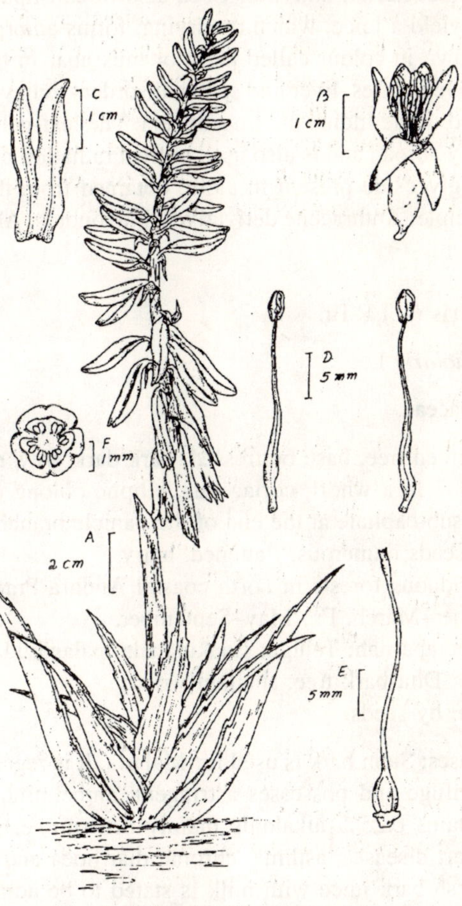

Fig. 13: *Aloe vera* (L.) Burm. f.
A. Habit; B. Flower; C. Perianth lobes (outer and inner); D. Stamens; E. Pistil; F. T.S. of ovary,

Common in drier parts of the state. It is cultivated in hedge rows.
Sanskrit: Kumari. Telugu: Chinna Kalabanda.
Trade name: Indian Aloe.
Propagation: by suckers.

Medicinal uses: Leaf pulp used to treat liver disorders, rheumatism, skin disorders, intestinal worms. It is also used as vanishing cream, emolient.

Fresh juice of the leaves is cathartic and cooling. It is used in eye troubles and spleen and liver ailments. It has been found useful for x-ray burns, dermatitis, cutaneous leishmaniasis and other disorders of skin. It is used for evacuation relief in the presence of anal fissures after roctonal operations. It is also used in stomach tumors, constipation, colic, worm infestation and worm infection.

Laxative, antibacterial, antiviral. Used also in constipation.

The leaves yield a juice, which on drying, forms amorphous, semi opaque, light to dark brown in colour called aloin or mussabar in trade. This is widely used as a purgative in piles, liver and spleen disorders, relieves inflammation and boils and is listed in the Indian pharmacopoeia. The leaf-gel has high demand in cosmetic industry abroad and is also used to treat radiation burns. The barrvaloin is the principal glycoside present in aloin, obtained from the leaf extract.

Active principle: Anthracene derivatives, Flavonoids, Aloeresins B,C and D. Aloin.

Alstonia scholaris (L.) R.Br.

Syn: *Echites scholaris* L.

Family: Apocynaceae.

Large evergreen tree, base buttressed, bark dark grey, nearly smooth, latex milky. Leaves 4–7 in a whorl, coriaceous, elliptic-oblong or obovate. Flowers greenish-white, sub-capitate at the end of the panicle branches. Follicles pendulous, cylindric, seeds numerous, flattened, hairy.

Rare in deciduous forests in north coastal Andhra Pradesh.
Fl.: December–March. Fr.: May–September.
Sanskrit: Saptaparnah; Telugu: Edakulapala, palagaruda.
Trade name: Dita bark tree, shaitam wood.
Propagation: by seeds.

Medicinal uses: Stem bark is used for Malaria. It is regarded as a bitter tonic and a mild febrifuge and possesses astringent, anthelmintic and galactogogue properties. Contains 0.25% alkaloids mainly echitamine. It is reported to be employed in heart diseases, asthma, chronic diarrhoea and to stop bleeding of wounds. The fresh bark juice with milk is stated to be administered in leprosy and dyspepsia. In Ayurvedic system, the drug is said to be useful in cancer-like conditions.

Alstonia venenata R.Br.

Family: Apocynaceae

Small tree, up to 4m tall. Leaves in whorls, oblong-lanceolate. Flowers white in terminal, sub-umbellate, dichotomous cymes. Fruit follicular, up to 12cm long, seeds with tuft of hairs.

Common in forests of Srikakulam, Visakhapatnam and Krishna districts.

Sanskrit: Rajadana. Telugu: Edakulapala.

Fl.: February–May, Fr.: May–July.

Propagation: By seeds.

Medicinal uses: The ripe fruits are stated to possess tonic and anthelmintic properties and are reported to be used as a remedy for impure blood, syphilis, insanity and epilepsy.

Chemistry: The plant is a rich source of indolealkaloids.

Alternanthera sessilis (L.) DC.

Syn: *Alternanthera triandra* Lam.

Gomphrena sessilis L.

Family: Amaranthaceae

Very variable prostrate to erect herb, up to 50cm tall, rooting from the basal nodes. Leaves lanceolate to spathulate. Flowers in axillary and terminal globose spikes, seeds orbicular.

Common weed of damp places along ponds and streams in most districts.

Sanskrit: Matsyaki, Lonika. Telugu: Ponagantikura

Prpagation: by vegetative means.

Medicinal uses: The herb is used as a galactogogue, cholagogue, abortifacient and febrifuge. It is also used for indigestion. The whole plant is a bitter, sweet, cooling and constipating. It is used to treat burning sensation, diarrhoea, leprosy, skin diseases, night blindness, dyspepsia and fever.

Amaranthus spinosus L.

Family: Amaranthaceae

Annual, erect armed herb, to 1m tall. Leaf axils with a pair of spines up to 2cm. Leaves ovate to lanceolate. Inflorescence axillary clusters and terminal spikes. Flowers numerous, sessile, unisexual. Fruit utricle, circumscissile. Seed orbicular, black.

Common in waste lands, fields and hedges near villages in most districts.

Fl. & Fr.: Throughout the year.

Sanskrit: Tandulya, Kataib, chaulai. Telugu: Mulla Thotakura.

Sanskrit: Matyaksi, Lonika. Telugu: Ponagantikura.
Trade name: Prickly Amaranth.
Propagation: by seeds.

Medicinal uses: The plant is regarded as an emmenagogue and galactogogue. It is used as a refrigerant, diuretic, purgative, enema for stomach troubles, piles, against cholera and also as a sudorific, it prevents vomiting.

Plant is a constituent of a herbal drug "Leu-cosol-H" which is found to be effective in Leucorrhoea.

Amorphophallus paeoniifolius (Denst.) Nicolson var. **campanulatus** (Decne) Sivadasan.

Syn: *A. campanulatus* non Blume ex Decne

Family: Araceae

A stout cormous herb, corm depressed globose, 25–30 cm across. Leaves 30–90 cm across, palmately lobed, segments pinnate or bipinnatisect, petiole dark green with pale blotches. Spathes broadly campanulate. Spadix globose, as long as spathe. Flowers monoecious, crowded in cylindric masses. Fruit berry, obovoid, red, 2–3-seeded.

Occasional in the hills of Eastern Ghats in Kurnool, East Godavari and north coastal districts.

Fl. & Fr: May–August.
Sanskrit: Suranah, Arsaghna, Telugu: Kanda gadda, Manchikanda
Trade name: Elephant Foot yam.

Medicinal uses: The corms are aperient, carminative and expectorant. Fresh corms are acrid, stimulant and increase appetite and taste. They are applied externally as an irritant to treat rheumatism. The corms are administered internally in the treatment of dysentery, piles and haemorrhoids. Corms are also used in cases of ear-aches, swelling of throat, pimples, intercostal neuralgia and fever. It is also useful in cases of elephantiasis, tumors, inflammations, constipation, seminal weakness, fatigue, anaemia and general disability

Amorphophallus sylvaticus (Roxb.) Kunth.

Syn: *Arum sylvaticum* Roxb.
Brachyspatha sylvatica (Roxb.) Schott.
Synantherias sylvaticus (Roxb.) Schott.

Family: Araceae

A cormous herb. Leaves large, 20–100 cm long, 10–60 cm broad, segmented, segments lanceolate. Flowers in spadix enclosed in spathe, spathe cream-coloured outside, maroon inside, male inflorescene 0.9–1.3 cm long, female 3–4 cm long.

Occasional in forests in moist localities.
Telugu: Adavikanda, vajrakanda, vanakanda.

Medicinal uses: The fruits and seeds are crushed into a paste, and used for toothache and bruises. They are also used in the treatment of glandular enlargement.

Ampelocissus latifolia (Roxb.) Planch.

Syn: *Vitis latifolia* Roxb.

Family: Vitaceae.

Tendrillar climbing shrubs, branches striate, tendrils forked, young shoots and inflorescence pubescent. Leaves orbicular, shallowly 3–5-lobed. Flowers reddish-green or reddish brown in leaf-opposed thyrsoid cymes. Fruit globose berry, edible, seeds usually 2, ellipsoid.
Occasional in forests.
Fl. & Fr.: June–September.
Telugu: Adavi Teega Draksha.
Propagation: by stem cuttings.

Medicinal uses: Tender leaves and roots used in dental troubles, indolent ulcers, Dysentery.

Anacardium occidentale L.

Family: Anacardiaceae

Trees, up to 8m tall, branchlets glabrous. Leaves simple, obovate, thick coriaceous. Flowers unisexual or polygamous, yellow streaked with pink in terminal panicles. Fruit nut, seated on a large pyriform fleshy body formed by the enlarged disc, seed reniform, testa membranous.
Cultivated in coastal districts, especially near the sea.
Fl. & Fr: January–May.
Sanskrit: Kajutaka. Telugu: Jidi mamidi
Trade name: Cashew nut.
Propagation: by seeds.

Medicinal uses: Kernel is antidote to irritant poisons. Also used in skin diseases and leprosy. Cashew bark is used for fevers, as a laxative and anthelminitic and to treat diabetes insipidus.

Anacyclus pyrethrum DC.

Syn: *Anacyclus officinarum* Heyne

Family: Asteraceae (Compositae).

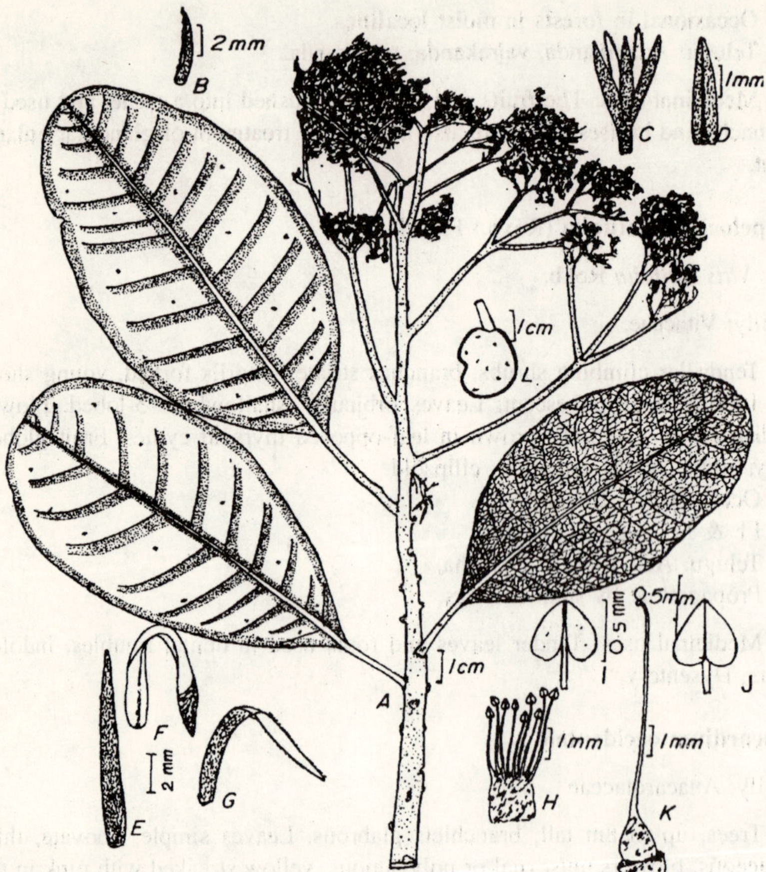

Fig. 14: *Ancardium occidentale* L.
A. Twig; B. Bract; C. Calyx; D. Sepal; E., F. & G. Petals; H. Androecium; I. & J. Stamens; K. Pistil.

Perennial procumbent herb, about 45cm tall, stems thickened, hollow. Leaves 2-pinnate. Flowers white tinged purple beneath in terminal heads. Fruits achene, winged.

Cultivated.
Telugu: Akkala karra
Trade name: Spanish pellitory.
Propagation: by seeds.

Medicinal uses: The roots have a slight aromatic smell and persistent, pungent taste and possess stimulant and rubifacient properties. The roots are used as a masticatory to allay toothache, for gargles and for the preparation of tooth powders. Roots are stimulant and rubifacient. The roots contain anacyclin, Pellitorine, hydrocarolin and Inulin.

Anamirta cocculus (L.) Wt.& Arn.

Syn: *Menispermum cocculus* L.

Family: Menispermaceae.

Large climbing shrubs. Leaves ovate-cordate. Flowers greenish in drooping panicles on old wood. Fruit druplets, white.

Rare in forests of Cuddapah.

Fl. & Fr.: October–April.

Sanskrit: Kakanasika. Telugu: Kakamari.

Trade name: Fish berry, Levant nut.

Propagation: by seeds.

Medicinal uses: Fruits and seeds are narcotic and are used to destroy lice, in curing skin diseases, scabies. They are also used as antidote for morphine and chloral poison.

Andrographis paniculata (Burm.f.) Wall. ex Nees

Syn: *Justicia paniculata* Burm.f.
Andrographis subspathulata Clarke

Family: Acanthaceae

Annual erect herb, up to 1.5 m tall, stem sharply quadrangular. Leaves lanceolate. Flowers white with maroon streaks on upper lip, in lax spreading axillary and terminal unilateral racemes. Fruit capsule, linear-oblong, acute at both ends, seeds numerous. Yellowish brown.

Common weed of plains and open forests all over the state.

Fl.& Fr.: July–December.

Sanskrit: Bhuninbah. Telugu: Nelavemu. English: King of bitters.

Medicinal uses: Whole plant is used as febrifuge and tonic. It is also used as alterative, antipyretic, antihistaminic, anthelmintic and cholagogue. It is extensively given as remedy for fevers and such other conelitions where chireyata (*Swertia chirayita*) is prescribed.

The herb is reported to possess astringent, anodyne, tonic and alexipharmic properties and is useful in dysentery, cholera, diabetes, influenza, bronchitis, swellings and itches, piles and gonorrhoea. The decoction of the plant is a blood purifier. It is used as a cure for torpid liver and jaundice. It forms the major constituent of the Ayurvedic drug Switradilepa' which is effective in treating vitiligo, a dermatological disease.

Active principle: Kalameghin and andrographolide 0.8–2.5%.

Andrographolide is Hepatoprotector.

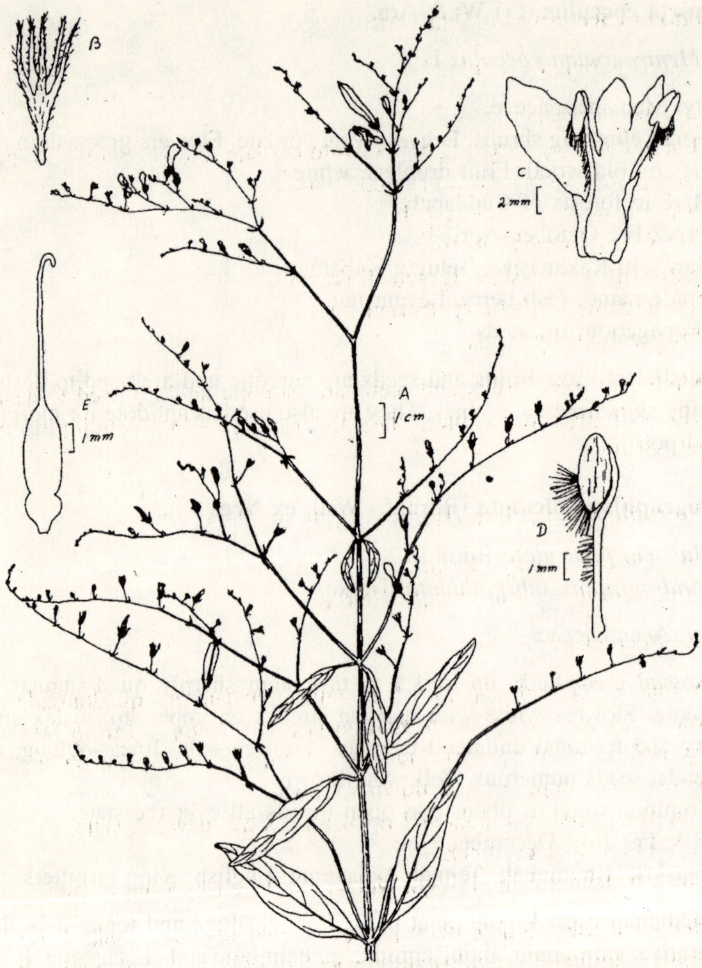

Fig. 15: *Andrographis paniculata* (Burm.f.) Wall ex Nees
A. Twig; B. Calyx; C. Corolla open; D. Stamen; E. Pistil.

Angiopteris evecta (Forst.) Hoff.

Family: Angiopteridaceae (Fern family)

Rhizomatous perennial fern. Stipes up to 150 cm X 5cm. Lamina deltoid; up to 230 x 140 cm, bipinnate, pinnae subopposite with 3 cm long stalk, oblong-lanceolate, pinnules subopposite, shortly stalked, oblong-lanceolate. Sori submarginal and ellipsoidal.

Common along shaded stream banks in thick forests of Visakhapatnam, East Godavari and Chittor districts.

Telugu: Kalojikanda

Medicinal uses: Rhizome extract mixed with milk, administered for chest pain and heart ailments-3 spoonfuls twice a day for 3 days.

Anisochilus carnosus Wall.

Syn: *Lavandula carnosa* L.f.

Family: Lamiaceae (Labiatae)

Aromatic erect herb, up to 50 cm tall, stem tetragonous. Leaves ovate, verrucose above and pubescent below. Flowers purple or pale violet in terminal long spikes. Fruit nut lets.

Common in moist rocky places on the hills.

Fl.& Fr.: October–March.

Sanskrit: Induparni. Telugu: Riddu-Rodda, Karpuravalli.

Trade name: Thick leaved lavender.

Fig. 16. *Anisochilus carnosus* (L.f) Wall. ex Benth.
A. Twig; B. Flower; C. Stamen; D. Pistil.

Propagation: by seeds.

Medicinal uses: The plant is stimulant, expectorant and diaphoretic. The juice of the fresh leaves is credited with cooling properties and is used to relieve urticaria associated with liver disorders and other allergic manifestations; mixed with sugar candy it is used as a domestic remedy for coughs and colds.

Anisomeles indica (L.) O.Kuntze

Syn: *Anisomeles ovata* R.Br.
Nepeta indica L.

Family: Lamiaceae (Labiatae).

An almost perennial, erect, pubescent undershrub, 1.5–2 m tall, stems and branches acutely tetragonous. Leaves ovate to obovate. Fruit nutlets, shining black when ripe.

Common all over the state in waste places and road sides.

Fl.& Fr.: October–January.

Fig. 17: *Anisomeles indica* R. Br.
A. Twig; B. Flower; C. Corolla; D. Stamen; E. Pistil.

Telugu: Ada beera
Propagation: by seeds.

Medicinal uses: The herb possesses aromatic, astringent, carminative and tonic properties and is employed as a cure in gastric catarrh and intermittent fevers. A decoction of the herb is given in convulsions and is applied in itches. The essential oil present in the herb is useful in uterine affections.

Anisomeles malabarica (L.) R.Br. ex Sims.

Family: Lamiaceae (Labiatae).

A silvery tomentose aromatic undershrub, up to 2m tall, stem and branches obtusely tetragonous. Leaves thick, lanceolate. Flowers purple in axillary and terminal spikes. Fruits nutlets.

Common all over in the state in waste lands and along roadsides.
Fl.& Fr.; Throughout the year.
Propagation: by seeds

Medicinal uses: The herb is reported to possess anti spasmodic, emmenagogue, diaphoretic, antipyretic and anti-periodic properties. A decoction of the herb is beneficial in rheumatic pains. The fresh juice of the leaves or an infusion is given to children in colic, dyspepsia and fever due to teething.

Annona muricata L.

Family: Annonaceae

Small evergreen tree, branchlets glabrous. Leaves obovate-oblong. Flowers yellowish, solitary, axillary. Fruit ovoid, dull green, tuberculate, seeds many, reddish brown.

Cultivated occcasionally.
Fl.& Fr.: April–october.
Telugu: Lakshmanaphalamu.
Trade name: Sour sop.
Propagation: by seeds.

Medicinal uses: The leaves and flowers are reported to be used in kidney troubles and as sudorific; the flowers and buds are used as a remedy for cough and in catarrh. A poultice of fresh leaves is stated to be applied to cicatrice, as a parasiticide. A decoction or infusion of the leaf is given in fever, dysentery and as a stomachic and anthelmintic. The root is said to be antispasmodic, antidysenteric, parasiticidal and piscicidal. The green bark is an astringent and a drastic purgative.

Annona reticulata L.

Family: Annonaceae.

Small trees. Leaves oblong-lanceolate. Flowers 2–4 on lateral pedicels. Fruit subglobose, yellow or yellowish-red when ripe.

Commonly cultivated in gardens in Nellore, Visakhapatnam and Nizamabad districts.

Fl.& Fr.: April–June

Sanskrit: Ramphal. Telugu: Ramaphalamu.

Trade name: Bullocks Heart of India.

Propagation: by seeds

Medicinal uses: The seed is reported to possess astringent and vermifugal properties and is used in diarrhoea and dysentery.

Annona squamosa L.

Family: Annonaceae

Shrubs or small trees, bark grey. Leaves simple, elliptic or oblong-lanceolate, glabrescent. Flowers greenish, solitary, on leaf-opposed peduncles or 2–4 in a cluster. Fruit a fleshy syncarp.

Common in forests of Telangana, occasional elsewhere. Also cultivated for its fruits.

Fl.& Fr.: April–November.

Sanskrit: Gandhagathra, Shubha. Telugu: Sitaphlamu.

Trade name: Custard apple.

Propagation: by seeds.

Medicinal uses: The root is employed in depression and spinal diseases, in asthma and fever. The bark is used in diarrhoea. The leaf is reported to possesses stimulant, antispasmodic and sudorific, anthelmintic and insecticidal properties. The crushed leaves are reported to be applied to the nostrils in hysteria and fits. The fruit possesses astringent, cooling, anti-scorbutic, febrifugal and pectoral properties. Fruit is a source of vitamin C.

Anodendron paniculatum (Roxb.) A.DC.

Syn: *Echites paniculata* Roxb.

Family: Apocynaceae

Liane; latex milky. Leaves opposite decussate, broadly oblong; Flowers yellow, in large, lax trichotomous cymes. Fruits paired, basally swollen, gradually tapering above. Seeds obovate with coma.

Rare in forests of Visakhapatnam and Warangal.

Fl.& Fr.: December–May.

Medicinal uses: The root crushed in water is used to aid parturition. It is reported to possess properties similar to ipecacuanha roots.

Chemistry: Wood contains 14 cardenolides called anodendrosides A-G.

Anthocephalus chinensis (Lam.) A. Rich ex Walp.

Syn: *Anthocephalus cadamba* (Roxb.) Miq.
Nauclea cadamba Roxb.

Family: Rubiaceae

Large deciduous tree, bark grey, vertically fissured. Leaves elliptic-oblong. Flowers greenish-yellow or orange in globular heads. Fruit is a pseudocarp, orange, fleshy, seeds many, minute, muricate.

Occasional along river banks and hill valleys in north coastal districts.

Fl.: May–July, Fr.: August–October.

Sanskrit: Kadamba; Telugu: Kadambamu.

Trade name: Kadam; Parvaty's tree

Medicinal uses: The stem bark is reported to possess tonic, astringent, febrifugal and anti diuretic properties, and is given in coughs. The juice of the bark forms a constituent of a compound to treat inflammation of the eye.

Aphanamixis polystachya (Wall.) Parker

Syn: *Aglaia polystachya* Wall.
Amoora rohituka (Roxb.) Wight & Arn.

Family: Meliaceae

A large evergreen tree, bark greyish brown. Leaves imparipinnate, leaflets 9–17, elliptic, oblong, glabrous. Flowers unisexual, dull white, male flowers in axillary panicles, female flowers in axillary or supra-axillary spikes; Fruit capsule, yellow when ripe, 3-valved, seeds 1–3, sub globose with red aril.

In semi evergreen forests of Visakhapatnam.

Fl.: August–September; Fr.: January–June.

Sanskrit: Rohituka, Saptavha, Telugu: Chemmaram, Chevamanu.

Trade name: Sohaga.

Medicinal uses: The bark is strongly astringent and is used in diseases of the liver and the spleen and for tumors and abdominal complaints. The pounded bark is used as a poultice in rheumatism. The seeds are credited with refrigerant, laxative and anthelmintic properties. They are used for curing ulcers and muscular pains.

Argemone mexicana L.

Family: Papaveraceae

Annual prickly herbs, stem greyish-white with yellow latex. Leaves both radical and cauline, white, spiny on margins and veins, sessile, amplexicaul, oblong-ovate, glaucous.

Fruit poricidal capsule, prickly, seeds blackish-brown.

Common weed in open wastelands and cultivated fields and road sides.

Sanskrit: Swarnakshiri. Telugu: Balurakkasi, Brahmadandi, Yerrikusuma.

Trade name: Prickly poppy, Mexican poppy.

Medicinal uses: The seeds are emetic and narcotic. They are reported to be used for diarrhoea and dysentery; but taken in large quantities, they are poisonous.

Chemistry: Contain Berberine.

Argyreia nervosa (Burm.f.) Bojer

Syn: *Argyreia speciosa* Sweet

Family: Convolvulaceae

A very large climbing shrub with stout, white-tomentose stems. Leaves broadly ovate or ovate-cordate, glabrous above, densely white-tomentose beneath. Flowers rose-purple, in axillary clustered cymes. Fruit ovoid berry.

Occasional in open forests in most districts.

Fl; & Fr: July–December.

Sanskrit: Samudrapalaka. Telugu: Samudrapala, Chandrapala.

Trade name: Elephant creeper.

Medicinal uses: In the indigenous system of medicine the plant is prescribed in gleet, gonorrhoea, strangury and chronic ulcers. The leaves are emollient and vesicant. They are used externally in the treatment of ring worm, eczema, itch and other skin diseases, and internally to cure boils, swellings etc. The leaves are also used as a local stimulant and rubifacient.

Chemistry: The leaves contain anthocyanin, quercetin, Kaempferol and its glycosides.

Aristolochia bracteolata Lam.

Syn: *Aristolochia bracteata* Retz.

Family: Aristolochiaceae

A slender, prostate or trailing herb. Leaves alternate, cordiform-reniform, glaucous below. Flowers dark purple, solitary. Fruit capsule, oblong-ellipsoid, ribbed, seeds cordate.

Common weed of cultivated fields and wastelands all over the state.
Fl. & Fr.: July–December.
Sanskrit: Kitamari, Dhumra-patra, Gridhrani. Telugu: Gadida-gadapaku.
Trade name: Wort Killer.

Medicinal uses: The plant is reputed in indigenous medicine for its bitter, purgative and anthelmintic properties. The root decoction is employed to expel round worms. It is reputed as an emmenagogue. The bruised leaves mixed with castor oil are applied externally in eczema. They are also applied to navels of children in colic.

Aristolochia indica L.

Family: Aristolochiaceae

Twining shrub, stem grooved. Leaves obovate-oblong. Flowers greenish-white, tube with globose inflated base, then bent at right angle and suddenly narrowed into a cylindric tube. Fruit capsule, oblong, 6-valved, opening from below upwards, seeds many, laterally winged.

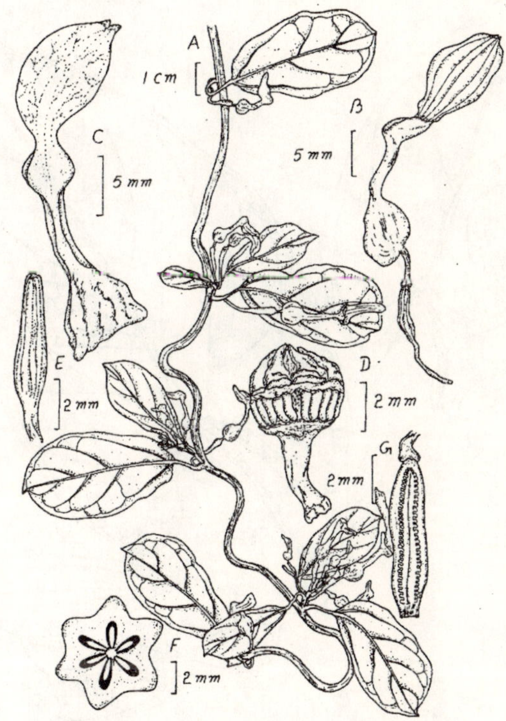

Fig. 18: *Aristolochia indica* L.
A. Twig; B. Flower; C. Perianth; D. Stylar column with stamens; E. Anther; F. T.S of ovary; G. L.S. of ovary.

Occasional in forests in all districts.
Fl.: & Fr.: September–March.
Sanskrit: Eswari, Sunanda; Telugu: Eswari.
Trade name: Serpent root plant, Indian birth wort.

Medicinal uses: The dried roots and rhizomes constitute an important drug, much esteemed as a gastric stimulant and bitter tonic.

The chief principle of the drug is aristolochic acid, though arostolic and p-coumaric acids also appear to contribute to the activities of the drug.

Aristolochia tagala Cham.

Syn: *Aristolochia acuminata* Roxb.
Aristolochia roxburghiana Klotzsch

Family: Aristolochiaceae

Climbing shrub, stem grooved. Leaves ovate or cordiform. Flowers dark

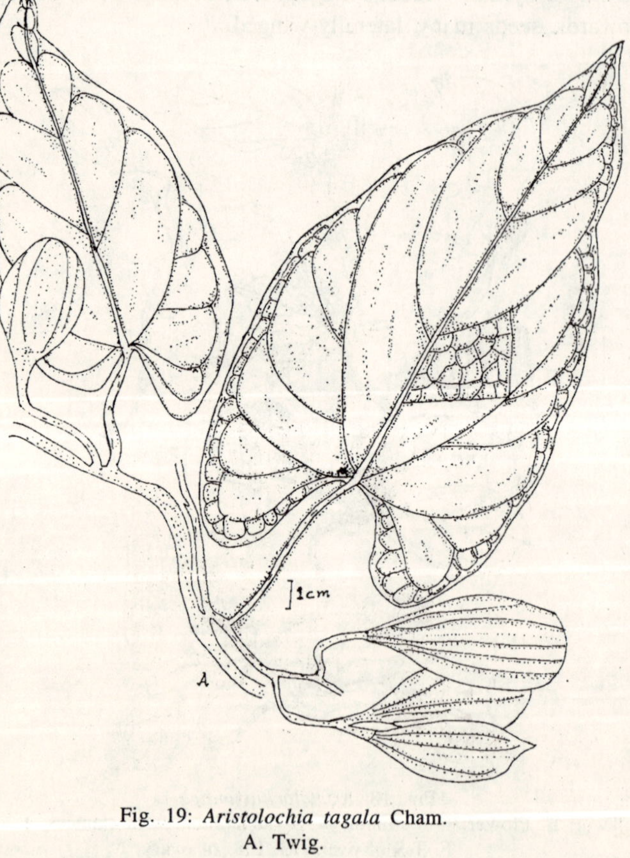

Fig. 19: *Aristolochia tagala* Cham.
A. Twig.

purple in axillary racemes. Fruit capsule, globose, seeds flat, thin and broadly winged.

Occasional at high elevations in Visakhapatnam district.
Fl. & Fr: September–December.
Telugu: Nalla Eeswari
Trade name: Heart leaved Indian birth wort.

Medicinal uses: Roots are carminative, tonic and for increasing breast milk.

Artabotrys hexapetalus (L.f) Bhandari

Syn: *Artabotrys odoratissimus* (Roxb.) R.Br.
Annona hexapetala L.f

Family: Annonaceae

Straggling shrub. Leaves oblong-lanceolate, glabrous. Flowers solitary or paired on leaf-opposed or terminal peduncle which is curved and hooked, dark green when young and yellow when mature.

Cultivated for its strongly fragrant flowers.
Fl.& Fr.: Round the year
Telugu: Tiga sampangi, Manoranjithamu
Propagation: by seeds.

Medicinal uses: The leaves are found to contain antifertility principle. A decoction of the leaves is given in cholera.

Artemisia absinthium L.

Family: Asteraceae (Compositae)

Herb to semi-shrub, 60–120 cm tall, Leaves long petioled, silky pubescent on both sufaces, the lower leaves abrupt pinnate and upper ones simple. Flowers in capitula, yellow, fruit achene. Cultivated.

Sanskrit: Indhana, Damar. Telugu: Machapatri.
Trade name: Worm wood, Madder wood
Propagation: by seeds.
Parts used: Flower heads, aerial shoots, leaves.

Medicinal uses: In chronic fevers, swellings, inflammation of liver, tonic, stimulant, rheumatism and worm infestation.

Active principle: Volatile oil, Sesquiterpenes.

Artemisia pallens Wall. ex DC.

Family: Asteraceae (Compositae)

Aromatic annual erect herb. Leaves much divided. Flowers yellow in head inflorescence.

Commercially cultivated.
Telugu: Davanamu
Propagation: by seeds.
Parts used: Leaves.

Medicinal uses: Anthelmintic, tonic, anti-pyretic.

Artocarpus heterophyllus Lam.

Syn: *Artocarpus integrifolius* auct. non L.f.

Family: Moraceae

A large evergreen tree, bark black mottled green, smooth. Leaves oblong, coriaceous, dark green. Inflorescence cauliflorous and also from the main branches, tender heads enclosed by glabrous leafy scales which fall off leaving annular scars at base. Fruit very large, elliptic-oblong, covered with spinous projections, seeds smooth, ovoid.

Cultivated all over the state, rare in evergreen forests.
Fl. & Fr.: January–July.
Sanskrit: Panasah. Telugu: Panasa
Trade name: Jack fruit tree.

Medicinal uses: The unripe fruit is acrid, astringent, carminative and tonic. The ripe fruit is laxative and cooling and useful in biliousness. The seeds are diuretic. The leaves are used in skin diseases. Ash of the leaves is useful in healing ulcers. The root is said to be useful in skin diseases, asthma and diarrhoea.

Asclepias curassavica L.

Family: Asclepiadaceae

Erect herb, up to 90 cm tall. Leaves decussate, lanceolate. Flowers bright orange red, in sub-terminal umbels. Fruit follicles, seeds winged, with silky coma.

In waste places and road sides, often cultivated as ornamental.
Fl.& Fr.: Throughout the year.
Sanskrit: Kakatundi Telugu: Jilledu mandara.
Trade name: Blood flower, False Ipecac.
Propagation: by seeds.

Medicinal uses: All parts of the plant are used in the indigenous system of medicine. The root is emetic, cathartic and used as an astringent and a remedy in piles and gonorrhoea. The juice of the leaves is antidysenteric, anthelmintic, sudorific and is also used against haemorrhages and gonorrhoea. The fresh or dried and pulverised leaf is reported to be used against cancer. The whole plant is considered emetic, styptic and purgative.

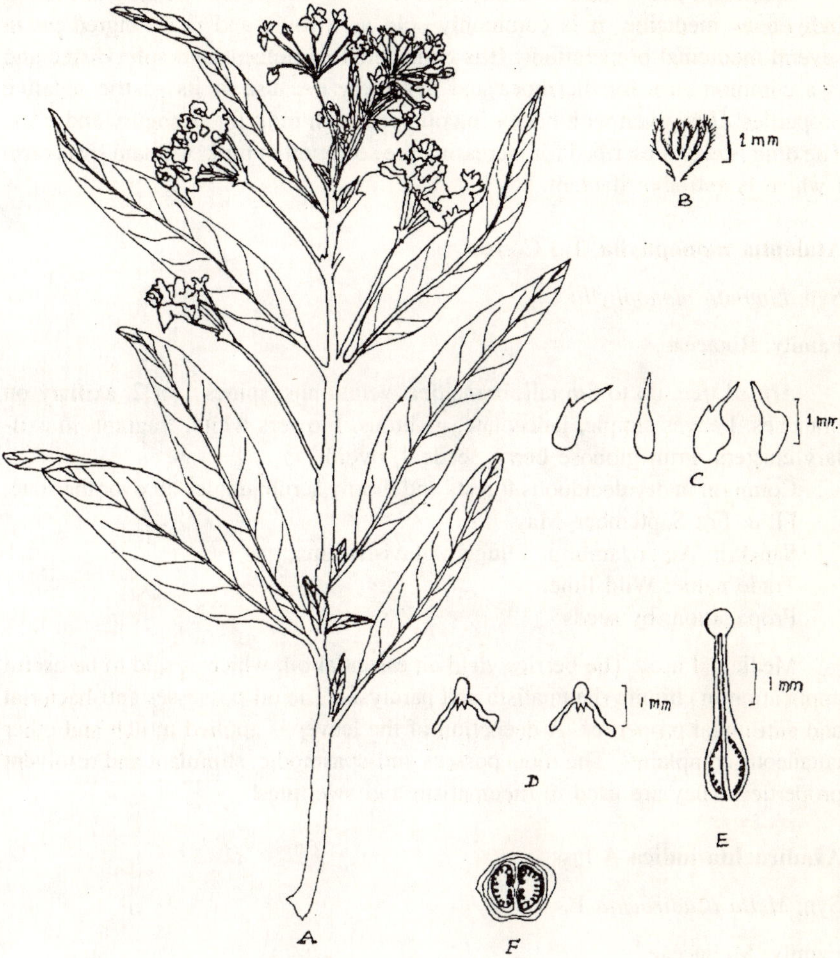

Fig. 20: *Asclepias curassavica* L.
A. Twig; B. Calyx split open; C. Corolla; D. Pollinia; E. Pistil; F. T.S. of ovary.

Asparagus racemosus Willd.

Family: Liliaceae

Perennial armed climbing shrub. Roots tuberous. Stem angular. Leaves scaly. Cladodes 2–6 in whorl, linear, falcate. Flowers white in racemes. Fruit berry, globose, red when ripe 3–6-seeded.

Occasional in forests in all districts.

Fl. & Fr.: September–October

Sanskrit: Satavari; Telugu: Pilligaddalu, challagaddalu.

Trade name: Wild Asparagus.

Medicinal uses: The root constitutes the drug shatawar. The drug is used in indigenous medicine. It is commonly sold as a tonic and is an ingredient in several medicinal preperations. It is considered demulcent and aphrodisiac and is a common cure for diarrhoea and dysentery because of its gastric sedative properties. It is given with honey in colic and with milk for strangury and piles. The drug is also prescribed for increasing the secretion of milk. Contain shatavarin I which is anti-abortifacient.

Atalantia monophylla (L.) Correa

Syn: *Limonia monophylla* L.

Family: Rutaceae.

Armed tree, up to 8m tall, branchlets velutionus, spines 1 or 2, axillary on branches. Leaves simple, lanceolate, glabrous. Flowers white, fragrant, in axillary clusters. Fruit globose berry, seeds 4, ovoid.

Common in dry deciduous forests and thorny scrub jungles all over the state.

Fl. & Fr.: September–May.

Sanskrit: Atavi-Jambira. Telugu: Adavi-Nimma.

Trade name: Wild lime.

Propagation: by seeds.

Medicinal uses: The berries yield an essential oil, which is said to be useful application in chronic rheumatism and paralysis. The oil possesses anti-bacterial and antifungal properties. A decoction of the leaves is applied in itch and other cutaneous complaints. The roots possess anti-spasmodic, stimulant and resolvent properties. They are used in rheumatism and swellings.

Azadirachta indica A.Juss.

Syn: *Melia azadirachta* L.

Family: Meliaceae

Medium sized tree, up to 15m tall, bark brown with vertical furrows. Leaves pinnate, leaflets 6–10 pairs, oblique. Flowers white in axillary panicles. Fruit one-seeded drupe, yellow when ripe, seeds ellipsoid, cotyledons thick, fleshy, oily.

Common all over Andhra Pradesh, both in plains and forests.

Fl.& Fr.: March–April, Frs: Ripen from May–July.

Sanskrit: Picumarda, Nimbah, Telugu: Vepa.

Trade name: Neem Tree, Margosa tree.

Propagation: The plant can be propagated through seeds in all types of soils. It grows best in places with 50–70 cm rainfall and temperature of 20–43° C.

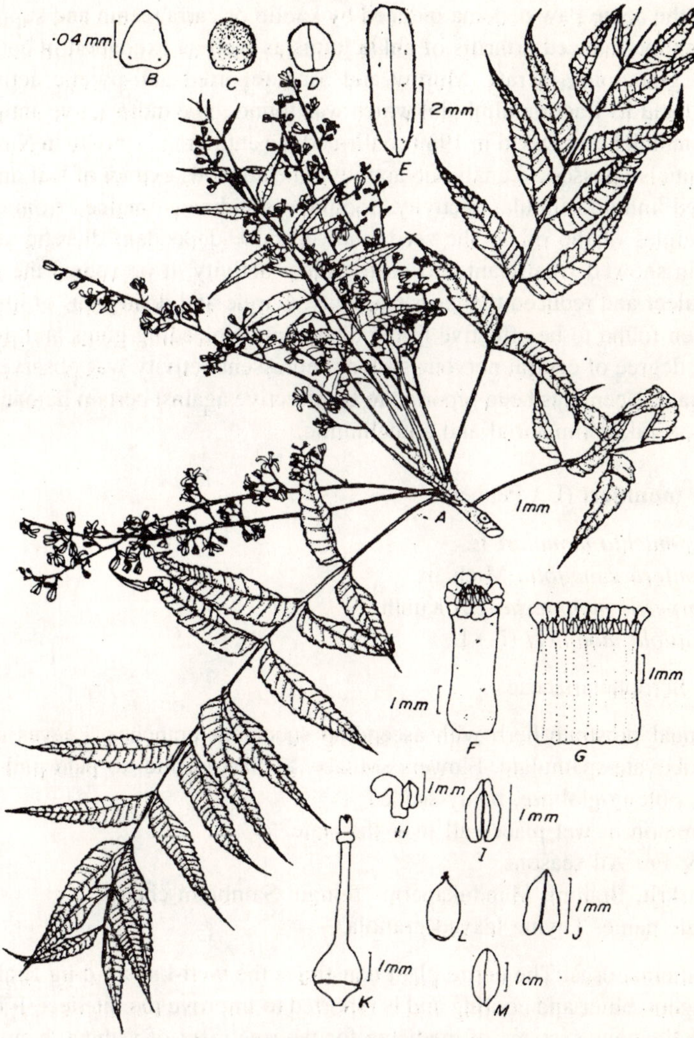

Fig. 21: *Azadirachta indica* A. Juss.
A. Twig; B-D. Sepals; E. Petal; F. & G. Staminal column whole and spread out; H. Staminal column lobe with anther; I. & J. Anthers; K. Pistil; L. Fruit; M. Seed.

Medicinal uses: Neem oil and its fractions, nimbidol and nimbidin have antipyretic activity. The plant also possessed analgesic activity. Sodium nimbidinate obtained from the bitter principle of the oil of the seed showed dose dependent diuretic activity.

Contains Gedunin which is antimalarial.

Seeds of neem contain Nimbidin ($C_{26}H_{34}O_6$), Nimbin ($C_{30}H_{36}O_9$) and Azadirachtin A ($C_{35}H_{44}O_{16}$). Nimbidin isolated from the seed oil significantly reduced the acute Paw oedema induced by kaolin or carrageenin and suppressed the formaline induced arthritis of ankle joints as well as exudation in cotton oil induced granuloma in rats. Murthy and Sirsi reported anti-pyretic activity in neem oil and its fraction nimbidol which was found to be more active antipyretic than acetanlid. Later David in 1969 confirmed the antipyretic activity in Nimbidin. The plant also possessed analgesic activity. The aqueous extract of leaf and bark possessed immunostimulant activity. Sodium nimbidinate obtained from the bitter principles of the oil of the seed, showed dose dependent diuretic activity. Nimbidin showed significant anti-gastric ulcer activity. It prevented the formation of ulcer and reduced the gastric activity. Gargle and dentrifices of nimbidin have been found to be effective in the treatment of bleeding gums and pyorrea. Varying degree of central nervous system depressent activity was observed with leaf extract. Neem has been reported to be effective against certain human fungi, bacteria, viral, antimalarial and anthelmintic.

Bacopa monnieri (L.) Pennel

Syn: *Lysimachia monnieri* L.
Moniera cuneifolia Michaux
Herpestis monnieria (L.) Kunth
Gratiola monnieri (L.) L.

Family: Scrophulariaceae

Annual prostrate herb with ascending succulent branches. Leaves sessile, fleshy, obovate-spathulate. Flowers solitary, axillary, white, to pale pink. Fruit capsule, oblong-globose, many-seeded.

Common in wet places all over the state.

Fl.& Fr.: All seasons.

Sanskrit: Brahmi, Mandukaparni: Telugu: Sambrani chettu.

Trade name: Thyme leaved gratiola.

Medicinal uses: The entire plant constitues the well-known drug Brahmi. It is astringent, bitter and cooling and is reported to improve the intellect. It is used in the indigenous systems of medicine for the treatment of asthma, hoarseness, insanity, epilepsy, and as a potent nerve tonic, cardiotonic and diuretic.

Bacosides-improve memory.

Balanites aegyptiaca (L.) Del.

Syn: *Balanites roxburghii* Planch.

Family: Balanitaceae.

Small armed tree, up to 4m tall, branches terete, striate, thorns up to 2cm long. Leaves 2-foliolate, leaflets elliptic-ovate or oblong, Flowers whitish, axillary or pseudo-axillary in pseudo umbels. Fruit drupe, ashy green, elliptic-oblong or ovoid.

Common in dry forests.

Fl.& Fr.: September–July.

Sanskrit: Ingudi, Angaravriksha. Telugu: Gara.

Trade name: Hingan, Ingudi.

Medicinal uses: Bark and fruit used to treat cold. Seeds used to treat whooping cough and colic. Pulp of ripe fruits used as a remedy for skin diseases.

Active principle: Diosgenin, isolated from seeds and from pericarp, is used in contraceptives.

Baliospermum montanum (Willd.) Muell.-Arg.

Family: Euphorbiaceae

Erect undershrubs, up to 2m tall. Leaves ovate-elliptic. Flowers greenish in axillary racemes, unisexual. Fruit capsule, ovoid, seeds ovoid, caruncled with fleshy endosperm and cotyledons, yielding oil.

Common along roadsides, fields in Telangana districts, occasional in north coastal districts.

Fig. 22: *Baliospermum motanum* (Willd.) Muell.-Arg.
A. & B. Twigs; C. Flower; D. Androecium; E. Stamens; F. Flower; G. Bract; H. Tepals; I. Pistil.

Fl.& Fr.: October–February
Sanskrit: Danti, Nikumbha, Udumbaraparni.
Telugu: Kondaamudamu, Adavi amudamu.
Trade name: Wild castor, Red physic nut.

Medicinal uses: The powdered seeds are used as a drastic purgative. Externally they are used as a stimulant and rubifacient. In Ayurvedic system of medicine the root is considered pungent, anthelmintic, diuretic and useful in skin diseases, piles, wounds and enlarged spleen.

Bambusa arundinacea (Retz.) Willd.

Syn: *Bambos arundinacea* Retz.
Bambusa bambos (L.) Voss
Arundo bambos L.

Family: Poaceae (Gramineae)

Large thorny shrub, culms thick, erect, 9–30m tall, internodes hollow. Leaf sheaths hairy, blade trigonous, broad. Inflorescence a pseudopanicle of spiklets, clustered in heads, spicate branchlets.

Occasional in forests in most districts.
Fl.& Fr.: Very infrequent
Sanskrit: Vamsah, Tawaksiri. Telugu: Veduru.
Trade name: Thorny bamboo.

Medicinal uses: The leaves are emmenagogue, anthelmintic, astringent and febrifuge. The stems and leaves are used in the Ayurvedic system of medicine as blood-purifier, in leucoderma and inflammatory conditions.

Barleria prionitis L.

Family: Acanthaceae.

A spinous bushy undershrub, about 1m tall, axillary spines 2–4, white, young shoots pubescent. Leaves elliptic-ovate, glabrescent. Flowers subsessile, solitary or in cymose clusters in lower axils. Fruit capsule, ovoid.

In waste lands and road sides.
Fl.& Fr.: September–January.
Sanskrit: Bindi, Mridukanta, Vajradanti. Telugu: Mullugorinta.
Trade name: Common yellow nail Dye.

Medicinal uses: The plant has antiseptic properties. The roots are used as a febrifuge. As a decoction they are employed as a mouth wash to relieve toothache and as a paste they are applied over boils and glandular swellings. The dried bark is given in whooping cough. Fresh juice of the bark is diaphoretic and expectorent and is given in anasarco. The leaf juice is applied to lacerated soles of feet in wet season, with coconut oil it is applied on the face for pimples. The fresh leaves are crushed and tied on the wounds caused by the sharp edged tools. They are also used for rheumatic pains and itch.

Barringtonia acutangula (L.) Gaertn.

Syn: *Eugenia acutangula* L.

Family: Barringtoniaceae-Lecythidaceae.

Evergreen tree, up to 10m tall, trunk short, bark thick, dark grey, distinctly furrowed on old trees, branchlets densely pubescent. Leaves thin-coriaceous, oblanceolate-ovate. Flowers pale pink or red in axillary, pendulous, spicate racemes. Fruit berry, ovoid with persistent calyx tube. Seed solitary, ovoid.

Common in plains along streams and swamp forests.

Fl. April–June, Fr.: December–January.

Telugu: Kanapa, Kurpa

Trade name: Indian oak, Small Indian Oak.

Medicinal uses: The fruit is bitter, acrid, anthelmintic and is prescribed in gingivitis as an astringent and tonic. It is also used as an emetic and expectorant. The leaf juice is given in diarrhoea. The root is bitter and possesses properties similar to those of cinchona.

Basella alba L.

Family: Basellaceae.

Trailing or twining vine. Leaves cordate, succulent. Flowers on fleshy spikes. Fruit berry, black when ripe

Cultivated.

Sanskrit: Upodika. Telugu: Bacchali.

Trade name: Indian spinach.

Propagation: Vegetative propagation with stem cuttings.

Medicinal uses: The herb is used as cooling medicine in digestive disorders. It is reported to be beneficial in fistulae, pustules and inflammatory tumors. A poultice of the leaves is applied to sores, warts and pimples. The leaf juice possesses demulcent, laxative and diuretic properties and is reported to be useful in gonorrhoea and in catarrh.

Bauhinia racemosa Lam.

Syn: *Piliostigma racemosa* (Lam.) Benth.

Family: Caesalpiniaceae.

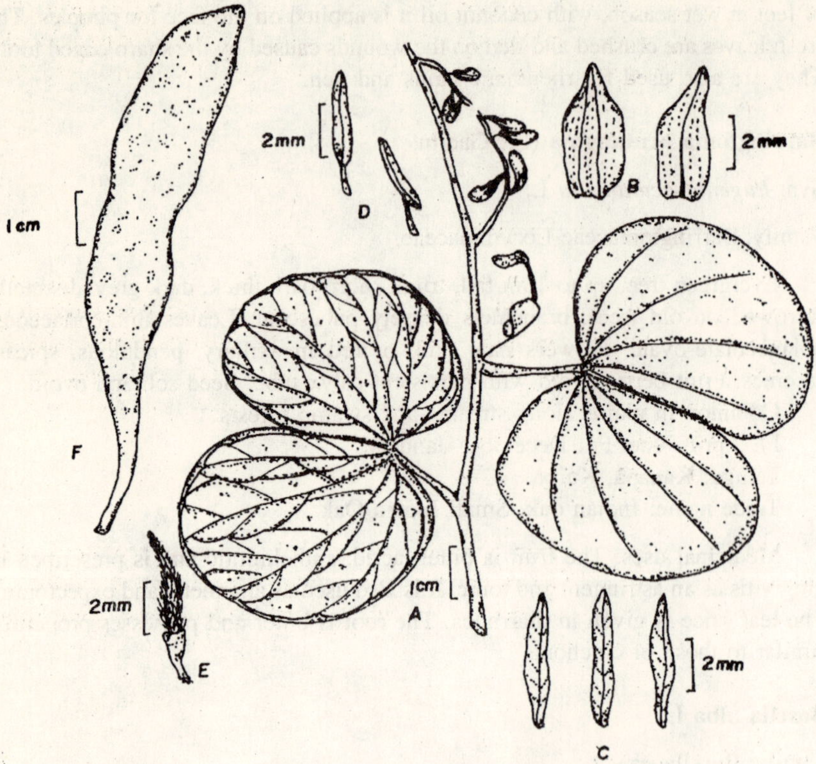

Fig. 23: *Bauhinia racemosa* Lam.
A. Twig; B. Sepals; C. Petals; D. Stamens; F. Pod.

Trees, about 10m tall, branchlets densely tomentose. Leaves ovate-orbicular, coriaceous, bilobed, leaflets connate to two thirds of their length. Flowers creamy yellow in terminal and leaf-opposed racemes. Pods linear-oblong, woody, slightly curved, dark brown, indehiscent, seeds about 15, flattened, ovoid.

Common in dry decidious forests.

Fl. & Fr.: March–November.

Sanskrit: Ashamantaka, Yamalapatraka. Telugu: Are chettu.

Trade name: Apta, Sweta kanchanara.

Medicinal uses: Bark and leaves are used to treat diarrhoea, dysentery and urinary discharges. Decoction of leaves used to treat malaria.

Bauhinia variegata L.

Family: Leguminosae (Fabaceae)

Deciduous tree, up to 13m tall, bark grey, longitudinally cracked. Leaves ovate, bilobed, leaflets connate about two thirds, thin coraceous. Flowers large,

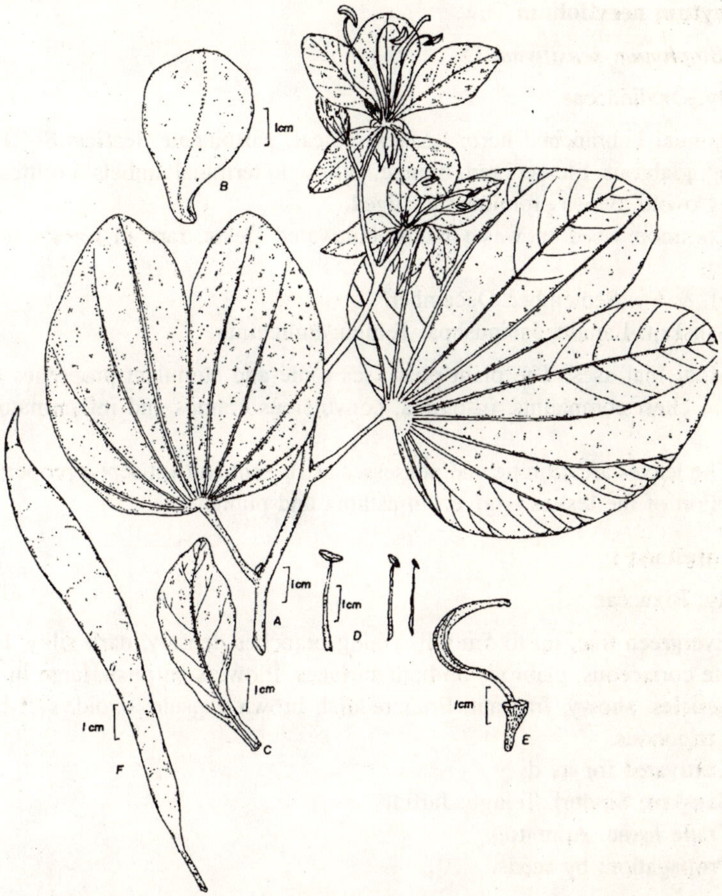

Fig. 24: *Bauhinia variegata* L.
A. Twig; B. Sepal; C. Petal; D. Stamens; E. Pistil; F. Pod.

white or pink or purple with one petal variegated yellow in short racemes. Fruit linear-oblong pod, brown, flat, dehiscent.

Occasional in forest, often cultivated.

Fl.& Fr.: February–May.

Sanskrit: Kavidara. Telugu: Devakanchanamu.

Trade name: Mountain Ebony.

Propagation: by seeds.

Medicinal uses: The root is carminative and is used in dyspepsia and flatulence. A decoction of the root is reported to prevent obesity. The bark is astringent, tonic and anthelmintic. It is useful in scrofula and skin diseases. It is also used for ulcers and leprosy. The dried flower buds are used for the treatment of diarrhoea and dysentery.

Biophytum nervifolium Thw.

Syn: *Biophytum sensitivum* DC.

Family: Oxalidaceae

Annual unbranched herb. Leaves radical, paripinnate, leaflets 8–10 pairs, oblong, glabrous. Flowers red, orange, yellow in terminal umbels. Fruit capsule, oblong-ovoid. Seeds prominently ridged.

Common weed in waste lands, cultivated fields, rare in forests in most districts.

Fl. & Fr.: September–December.

Sanskrit: Lajjalu, Jalapushpa. Telugu: Pulichinta.

Medicinal uses: The plant possesses tonic and stimulant properties and is used in chest complaints, insomnia, convulsious cramps and inflammatory tumors.

The leaves are diuretic and possesses astringent and antiseptic properties. A decoction of the leaves is given in asthma and phthisis.

Bixa orellana L.

Family: Bixaceae.

Evergreen tree, up to 5m tall, young branches densely, dark silky. Leaves cordate coriaceous, glabrous on both surfaces. Flowers purplish, large in terminal panicles, showy, fragrant. Fruit reddish brown capsule, ovoid, soft bristly, seeds trigonous.

Cultivated for its dye.

Sanskrit: Sinduri. Telugu: Jaffra.

Trade name: Annotto.

Propagation: by seeds.

Medicinal uses: The root bark is reportedly antiperiodic and is used in the treatment of fevers. The decoction of leaves is used as a gargle for sore throat. An infusion is given in jaundice and in dysentery. The pulp surrouding the seed is widely employed in medicine as a haemostatic, astringent, antidysenteric, diuretic, laxative, febrifuge and digestive. It is also prescribed in the treatment of epilepsy and skin diseases.

Boerhavia diffusa L.

Syn: *Boerhavia repens* L.

Family: Nyctaginaceae.

Annual, diffuse, ascending much branched herb. Leaves opposite, variable in size and shape. Flowers pink in axillary and terminal umbellate panicles. Fruit anthocarp, glandular.

Common in waste lands and roadsides all over Andhra pradesh.
Fl. & Fr.: Throughout the year.
Sanskrit: Punarnava, Sothagni. Telugu: Atika mamidi.
Trade name: Hog weed, Pig weed.
Propagation: by seeds.

Medicinal uses: Punarnava is diuretic, aphrodisiac, diaphoretic, emetic, expectorant, febrifuge and laxative, useful in nephrotic syndrome, chronic oedema and liver disorders. It is also used for asthma, anaemia, inflammations, leucorrhoea, ophthalmia, myalgia, constipation, cough, rheumatic and gouty complaints, kalazar, chronic peritonial conditions and heart diseases.
Active principle: Root contains alkaloid punarnavine (0.04%).

Bombax ceiba L.

Syn: *Bombax malabaricum* DC.
Salmalia malabarica (DC.) Schott.& Endl.

Family: Bombacaceae.

Tall deciduous tree, up to 30m tall, armed with prickles, bark greyish, smooth. Leaves crowded at the ends of branches, digitately 5–7-foliolate, glabrous. Flowers large, solitary or paired, bright crimson. Fruit capsule, oblong. Seeds many, enveloped by dense white silky floss.
Along streams in deciduous forests, often cultivated.
Fl.: February–April, Fr.: ripen from March to May.
Sanskrit: Nirgandha pushpi, Panchaparni. Telugu; Buruga.
Trade name: Red silk cotton Tree.
Propagation: by seeds.

Medicinal uses: The roots are used in dysentery, influenza, burning sensation, haemorrhoids and blood impurities. The bark is used for healing wounds. A paste of the bark is reported to be used against skin eruptions. Flowers are used in skin troubles. Young fruits are reported to be used to treat chronic inflammations and ulceration of bladder and kidney. The bark exudes a gum known as smul gum or mocharus, which is credited with astringent, tonic and demulcent properties and is used for dysentery, haemoptysis, in pulmonary tuberculosis, influenza and menorrhagia. Seeds are used for treating gonorrhoea and chronic cystitis.

Borassus flabellifer Murr.

Syn: *Borassus flabelliformis* Murr.

Family: Arecaceae (Palmae).

A tree with tall stem, up to 40m tall. Leaves simple, palmately fan shaped, plicate, petiole stout, spinous. Flowers unisexual, dioecious (male and female plants separate). Fruit a large subglobose drupe, yellow when ripe with 1–3 fibrous pyrenes.

Common in coastal regions, occasional inland.

Fl. & Fr.: February–April.

Sanskrit: Tala, Tadarumah. Telugu: Tadichettu.

Trade name: Palmyra palm.

Propagation: by seeds.

Medicinal uses: The root is considered cooling, restorative, diuretic and anthelmintic. It is employed as a cure for gonorrhoea. The roots are also useful in hyperdipsia, burning sensation and inflammation. Fruits are useful in dyspepsia, flatulence, colic, constipation, intestinal worms, leprosy, skin diseases and haemorrhages.

Boswellia ovalifoliolata Bal.& Henry

Family: Burseraceae.

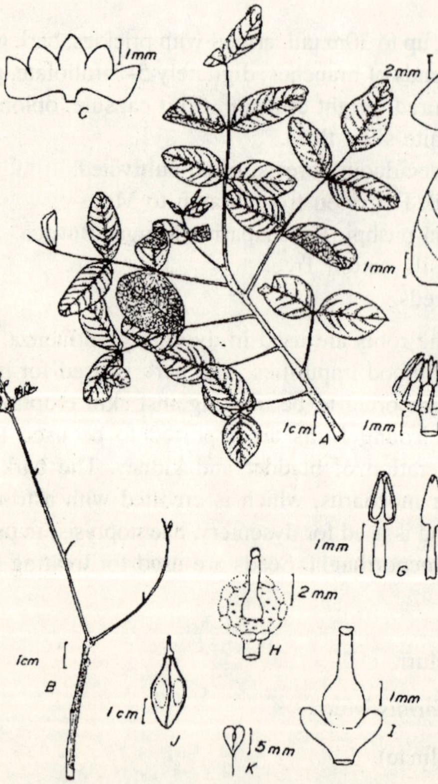

Fig. 25: *Boswellia ovalifoliolata* Bal. & Henry
A. Twig; B. Inflorescence branch; C. Calyx; D. Corolla; E. Petal; F. Androcium; G. Stamens; H. Pistil with disc; I. Pistil; J. Capsule; K. Seed.

Moderate sized deciduous tree, up to 10m tall, bark greyish, smooth, exfoliating in thin flakes. Leaves imparpinnate, leaflets 5–13, ovate or elliptic-ovate, glabrous. Flowers greenish white in monochasial cymes on much branched panicles. Fruit drupe, trigonous, pyrenes 3, bony.

Rare in Seshachalam hills in Chittoor and Cuddapah districts. Endemic to this region.

Fl.: December–February. Fr.: April–June.

Telugu: Konda sambrani.

Medicinal uses: Stem and bark are used to cure hydrocoel, inflammation and swellings, stomach ulcers.

Boswellia serrata Roxb. ex Colebr.

Syn: *Boswellia glabra* Roxb.

Family: Burseraceae.

Moderate seized deciduous tree, bark greenish, ashy grey, exfoliating in thin flakes. Leaves apically clustered, imparipinnate, leaflets 14–26, thin coriaceous, oblong-lanceolate. Flowers pinkinsh white in axillary panicles. Fruit drupe, trigonous, brown, pyrenes 3, each 1-seeded, seeds heart-shaped.

Common on dry hill slopes of dry deciduous forests all over the state.

Fl.: February–March, Fr.: May–June.

Sanskrit: Palasha, Kundru, Ashramurthi. Telugu: Guggilam, Anduga.

Trade name: White Dammer.

Propagation: by seeds.

Medicinal uses: Bark is used against diarrhoea and skin diseases. Gum is considered diuretic and stomachic and also used for the treatment of diarrhoea, dysentery, pulmonary and cutaneous troubles. A special extract (H15) of the gum resin is effective in the treatment of rheumatoid arthritis. Gum exudate of *Boswellia serrata* (known as Gum olibanum), which is sweet and bitter is credited with astringent, stimulant, expectorant, diuretic, diaphoretic, antipyretic, stomachic, emmenagogue, ecbolic and antiseptic properties. It is reported to be useful in ulcers, tumors, goitre, cystic breast, diarrhoea, dysentery, piles, asthma, bronchitis, chronic laryngitis, jaundice, syphilitic and skin diseases. The non phenolic fraction of the oleo-gum resin possess antitumor, sedative and analgesic activity.

Breynia vitis-idaea (Burm.f.) Fischer

Syn: *Rhamnus vitis-idaea Burm*.f.
Breynia rhamnoides (Retz.) Muell.-Arg.
Phyllanthus rhamnoides Retz.

Family: Euphorbiaceae.

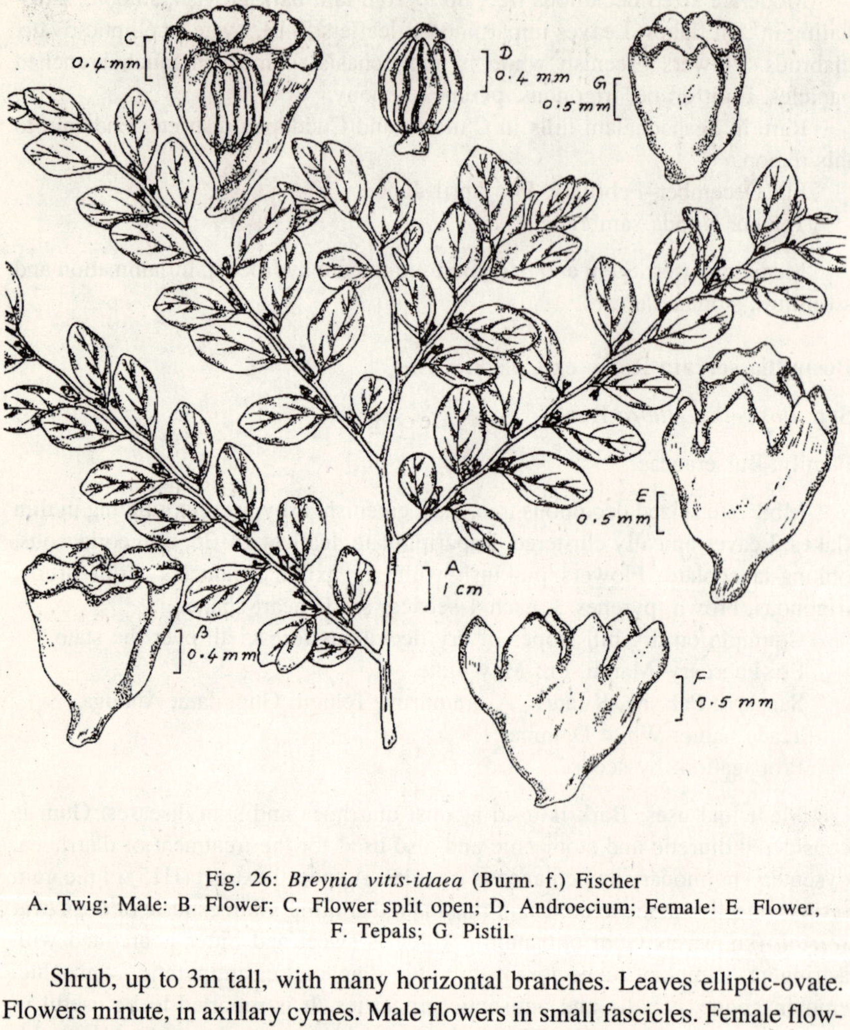

Fig. 26: *Breynia vitis-idaea* (Burm. f.) Fischer
A. Twig; Male: B. Flower; C. Flower split open; D. Androecium; Female: E. Flower, F. Tepals; G. Pistil.

Shrub, up to 3m tall, with many horizontal branches. Leaves elliptic-ovate. Flowers minute, in axillary cymes. Male flowers in small fascicles. Female flowers solitary. Fruit globose capsule.

Common in decidious forests.

Fl. & Fr.: September–January.

Telugu: Errapurugudu, Erraballi, Pandu Pachari.

Trade name: Coral Berry Tree.

Propagation: by seeds.

Medicinal uses: A decoction of the root is employed as mouth wash for tooth ache. The leaf juice is given to women after child birth to prevent haemorrhage.

Chemistry: The root contains b-sitosterol.

Buchanania lanzan Spreng.

Syn: *Buchanania latifolia* Roxb.

Family: Anacardiaceae.

Evergreen tree, up to 10m tall, bark dark grey, fissured. Leaves broadly oblong-ovate, glabrescent above, slightly villous beneath. Flowers creamish in axillary and terminal villous panicles. Fruit globose drupe, black, seeds 1 or 2

Occasional in deciduous forests.

Fl.& Fr.: January–May.

Sanskrit: Tapasya-priya. Telugu: Sarapappu.

Trade name: Almondette tree, Cheronjee.

Propagation: by seeds.

Medicinal uses: The stem exudes a pale or dark coloured gum, which is used in diarrhoea and inter coastal pains. The leaves are reported to be cooling, digestive, expectorant, purgative, depurative and aphrodisiac and are useful in hyperdipsia, burning sensation, cough, bronchitis, dyspepsia, flatulence, constipation, leprosy, skin diseases and seminal weakness. The fruits are useful in treating leprosy, skin diseases, gleet, inflammations, nervous debility, burning sensation, cardiac debility, abdominal disorders, constipation, stragury, cough, asthma, seminal weakness, fever, emciation, ulcers, general debility and as a laxative.

Butea monosperma (Lam.) Taub.

Syn: *Erythrina monosperma* Lam.
Butea frondosa Koen.ex Roxb.

Family: Leguminosae (Fabaceae).

Small to medium sized deciduous tree, trunk crooked and irregular, bark grey or dark brown, rough, branchlets tawny tomentose. Leaves trifoliolate, lateral leaflets rhomboid, terminal leaflet ovate. Flowers flame coloured or scarlet. The tree is leafless when flowering. Fruit pod, velvety brown, base wing like, indehiscent with out seeds. Seeds compressed, brownish.

Common in Telangana forests, occasional in Rayalaseema and coastal districts.

Sanskrit: Palasah. Telugu: Moduga.

Trade name: Flame of the forest, Bengal king.

Medicinal uses: The bark is reported to possess astringent, bitter, pungent, alterative, aphrodisiac and anthelmintic properties. The roots are useful in elephantiasis and in curing night blindness and other defects of sight.

Contains palasonin which is anti-helminthic.

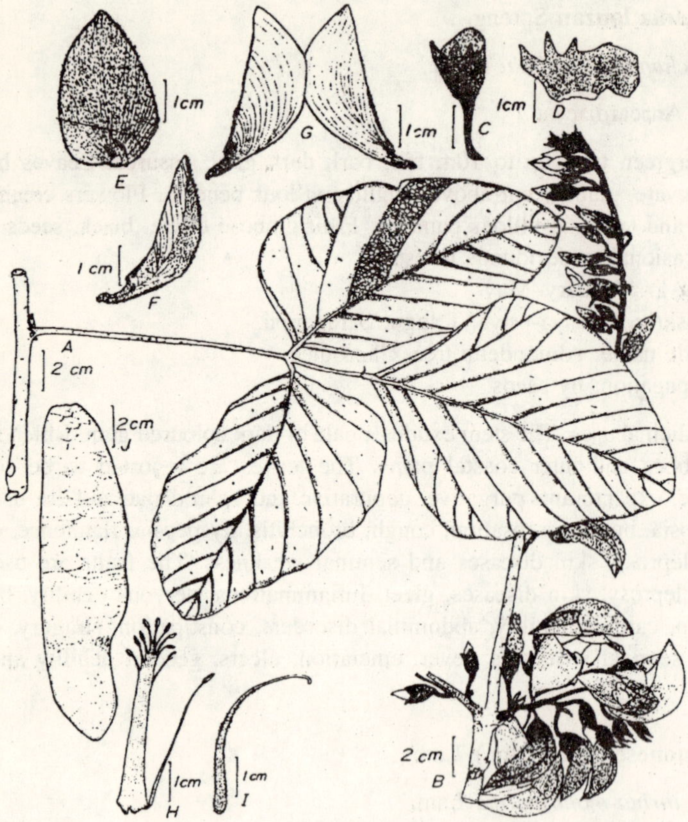

Fig. 27: *Butea monosperma* (Lam.)Taub.
A. Vegetative twig; B. Flowering twig; C. Calyx; D. Calyx split open; E. Standard petal; F. Wing petal; G. Keel petal; H. Staminal column split open; I. Pistil; J. Pod.

The flowers are reported to possess astringent, diuretic, depurative, aphrodisiac and tonic properties. They are used as an emmengogue and as poultice in ortitis and to reduce swellings for bruises and sprains. They are also effective in leprosy, leucorrhoea and gout.

The seeds are reported to possess aperient and rubefacient properties.

The leaves are credited with astringent, tonic, diuretic and aprodisiac properties. They are used to cure boils, pimples and tumerous haemorrhoids and are internally given in flatulent colic, worms and piles.

Butea superba Roxb.

Family: Fabaceae (Papilionaceae)

Large woody twiner. Leaves 3-foliolate, leaflets rhomboid or broadly obovate. Flowers orange red in racemes. Pods flat, tomentose.

Common in the interior deciduous forests.
Fl.& Fr.: March–May.
Sanskrit: Latha-palasha. Telugu: Tiga Moduga.
Propagation: by seeds.

Medicinal uses: The seeds are extensively used as sedative and anthelmintic. The decoction of seeds in considered emollient and is used topically for piles. The bark is used in tonics elixirs and in poultices.

Caesalpina bonduc (L.) Roxb.

Syn: *Guilandinia bonduc* L.
Caesalpinia crista L.
Caesalpinia bonducella Flem.

Family: Caesalpiniaceae

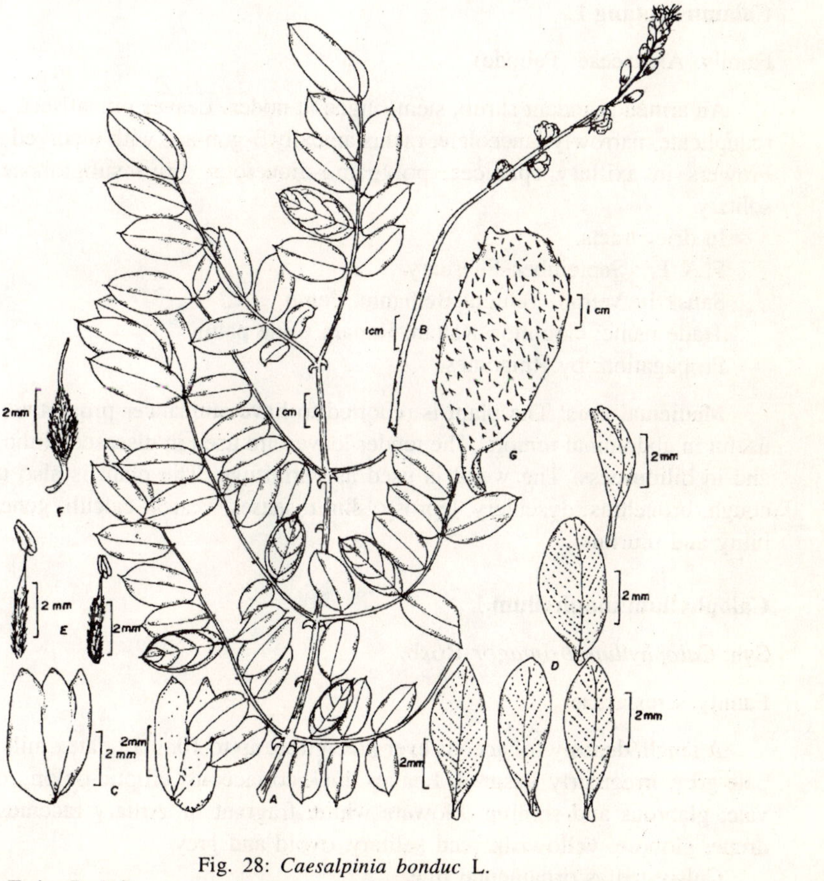

Fig. 28: *Caesalpinia bonduc* L.
A. Twig; B. Inflorescence; C. Sepals; D. Petals; E. Stamens; F. Pistil; G. Pod.

Armed straggler, prickles recurved, branchlets densely pubescent. Leaves bipinnate, pinnae 6–7 pairs, leaflets 6–8 pairs per pinna, elliptic-oblong. Flowers in axillary racemes, yellow. Fruit pod, oblong, strongly prickled, 1–2-seeded. Seed subglobose.

Occasional in plains in hedges and waste lands.

Fl.& Fr.: June–November.

Sanskrit: Kuberakshi. Telugu: Gacchakaya.

Medicinal uses: Leaves and seeds emmengogue and febrifuge. The seeds are bitter and show diuretic and antipyretic activity. They also proved efficaceous in diarrhoea. The seeds are also made into an ointment for treating hydrocele. As an infusion they are used for curing cerebral haemorrhage and infantile convulsions.

Active principle: Amorphous glycoside from seed kernel.

Calamus rotang L.

Family: Arecaceae (Palmae)

An armed scandent shrub, stem ringed at nodes. Leaves pinnatisect, leaflets reduplicate, narrowly lanceolate, rachis apically 3-gonous, with recurved spines. Flowers in axillary spadices, polygamo-dioecious. Fruit subglobose, seed solitary.

In drier tracts.

Fl.& Fr.: September–February.

Sanskrit: Vetrah. Telugu: Bethamu, Pemu.

Trade name: Common Rattan, Rattan, Cane palm.

Propagation: by Rhizomes.

Medicinal uses: The plant is reported to have anticancer properties, and is useful in abdominal tumors. The tender leaves are used in diseases of the blood, and in biliousness. The wood is used as vermifuge. The plant is also used in cough, bronchitis, dysentery, leprosy, skin diseases, vesical calculi, general debility and pruritus.

Calophyllum inophyllum L.

Syn: *Calophyllum brintagor* Roxb.

Family: Clusiaceae.

A small, densely foliaceous ever green tree, up to 10m tall, latex milky, bark pale-grey, irregularly fissured. Leaves thick-coriaceous, elliptic-oblong or obovate, glabrous and shining. Flowers white, fragrant in axillary racemes. Fruit drupe, globose, yellowish, seed solitary, ovoid and grey.

Cultivated as ornamental tree.

Fl.& Fr.: Throughout the year.
Sanskrit: Punnaga, Surangi. Telugu: Ponna chettu.
Trade name: Alexandrian Laurel.
Propagation: by seeds.

Medicinal uses: A paste of the kernals is applied to relieve painful joints. The oil from seeds is used as a stimulant embrocation in rheumatism and gout. The seed oil has a beneficial influence over the mucous membrane of the genito-urinary organs, and is, therefore used in the treatment of venereal diseases. The refined oil is intramuscularly injected to alleviate pain in leprosy. The oil cures scabies and other cutaneous diseases. The stem bark is astringent. The pounded bark is applied in orchitis and its juice is taken as purgative. The leaves are employed for inhalation in migraine and vertigo.

Calotropis gigantea (L.) R.Br.

Syn: *Asclepias gigantea* L.

Family: Asclepiadaceae.

Soft white woolly shrubs, up to 2m tall, latex milky. Leaves decussate, obovate-oblong, sparsely pubescent. Flowers purplish in axillary and terminal umbellate cymes. Fruits paired follicles, oblong, inflated. Seeds oblong with white silky coma.

Common weed of wastelands and cultivated fields.
Fl.& Fr.: Through out the year.
Sanskrit: Arkah. Telugu: Thella jilledu.
Trade name: Gigantic swallow wort.
Propagation: by seeds.

Medicinal uses: The root bark constitutes the drug, official in Indian pharmacopoeia. In small doses, the root bark is diaphoretic and expectorant. It acts as a mild stimulant, and is given with carminatives in dyspepsia. It is useful in leprosy and eczema. The powdered root bark gives relief in diarrhoea and dysentery. It is also given in cough and asthma, and as a febrifuge.

Active principle: Bitter resins akundarin. The latex contains the cardiac glycosides calotropin, gigantin and uscharin.

Calotropsis procera (Ait.) R. Br.

Syn: *Asclepias procera* Ait.

Family: Asclepiadaceae.

Erect shrub, latex milky. Leaves opposite decussate, obovate. Flowers purplish mauve, in axillary and terminal umbellate cymes. Fruit paired follicles, recurved. Seeds oblong with silky white coma.

Common weed in wastelands and cultivated fields.
Fl.& Fr.: Through out the year.
Sanskrit: Alarka. Telugu: Jilledu.
Trade name: Akund, swallow-wort.
Medicinal uses: Similar to that of *Calotropsis gigantea.*

Calycopteris floribunda Lam.

Family: Combretaceae.

Scandent shrubs, branches drooping, rusty brown-puberulous. Leaves alternate-subopposite, oblong or elliptic-oblanceolate, prominently gland dotted beneath. Flowers greenish white, in terminal and /or axillary, large fulvous panicles. Fruit ovoid-ellipsoid, 5-ribbed, crowned by persistent calyx.

Common in deciduous forests.
Fl. & Fr.: February–May.
Sanskrit: Susavi, Shwetadhataki. Telugu: Bontha.
Propagation: by seeds.

Medicinal uses: The leaves are bitter, stimulant and astringent; they are given in dysentery, colic, dyspepsia and fevers. They are also applied to sores and ulcers. The fruits are used along with other carminatives in jaundice. They are also applied with honey to treat ulcers.

Canavalia gladiata (Jacq.) DC.

Syn: *Dolichos gladiatus* Jacq.
Canavalia ensiformis auct.

Family: Fabaceae (Papilionaceae)

Stout, woody perennial or biennial twining shrubs, branches terete. Flowers purplish in axillary racemes. Fruit legume, thickened along both sutures, seeds 8–20, pink or red.

Two varieties, one wild (not edible) and another cultivated (edible) occur in the state.
Fl.& Fr.: September–March.
Sanskrit: Maha-Shibee. Telugu: Thamba, Thamma.
Trade name: Jack Bean, Sword Bean.
Propagation: by seeds.

Canna indica L.

Syn: *Canna orientalis* Roscoe

Family: Cannaceae.

Perennial erect herb, to 1.5m tall. Leaves elliptic-lanceolate, glabrous. Flowers scarlet, orange or yellow in terminal spikes. Fruit globose or ellipsoid capsule.

Grown as an ornamental.

Fl. & Fr.: Round the year.

Sanskrit: Devakuli. Telugu: Krishna Tamara.

Trade name: Indian Reed, Flowering reed.

Propagation: by Rhizomes.

Cannabis sativa L.

Family: Cannabinaceae.

Annual herb, 1–2m tall. Laeves opposite below, alternate above, 7–9-fid below, 3–5-fid above. Flowers unisexual, dioecious. Fruit globose achene.

Cultivated clandestinely.

Sanskrit: Bhang. Telugu: Ganjayi

Trade name: True Hemp, Bhanga, Ganja, Charas.

Medicinal uses: It is a narcotic plant. Ganja is prepared from the flowering and fruiting tops which are coated with a resinous exudate. Cannabis is recommended as a psychiatric aid in treating depression and high blood pressure. It is useful in treating glaucoma by reducing intra-occular hypertension. It is also prescribed in menstrual disorders and during labour pains.

Active principle: Cannabinol, pseudo cannabinol and cannabinin.

Bhang: Bhang consists of the dried, chrushed mature leaves and flower shoots of the male and/or female plants. The Euphoric principle develops in the plant only when it matures and flowers.

Ganja: It is prepared from the flowering and/or fruiting tops of *Cannabis sativa* which are coated with resinous exudate.

Charas or Hashish: It is a resinous matter collected from the leaves and flowering tops of the female plants.

Cultivation: *Cannabis sativa* requires a rich friable loamy soil. In summer the soil in the nursery is ploughed, well manured and laid into 30cm high ridges spaced 30 cm apart. The seeds are sown on a bright sunny day in August, the seed rate being 5–8kgs/ha. Seedlings of 15–30cm height are trasplanted in the field in rows 120 cm apart and 15–20cm between plants in each row. The field is given hand-weeding and interculturing to keep crop free of weeds. The crop is irrigated 4–6 times during the whole cropping season. The crop grows thick and bushy and reaches a height of 120–180cm by November. When the plants start flowering, only female plants are allowed to remain for the drug.

Canscora decussata Roem.& Schult.

Family: Gentianaceae.

Annual erect herb, up to 30cm tall, stem 4-winged. Leaves elliptic-lanceolate,3-nerved. Flowers white or pale pink in axillary or terminal dichasial cymes. Fruit obscurely 2-valved capsule. Seeds angular.

Occasional in moist places in most districts.
Fl. & Fr.: November–March.
Sanskrit: Sankhapushpi.

Medicinal uses: The herb is bitter, pungent and oleaginous and is regarded as laxative and tonic. It is reported to be used as a febrifuge. The fresh juice of the herb is prescribed in nervous debility. It is reported to promote memory power, cure leprosy, ulcers, oedoma, urinary disorders and snake bites.
Active principles: Xanthones.

Capparis zeylanica L.

Syn: *Capparis horrida* L.f.

Family: Capparaceae.

Large climbing shrubs with hooked spines, young parts rusty tomentose. Leaves ovate or elliptic, rusty tomentose when young, glabrous at maturity. Flowers yellowish white or white in supra-axillary peduncles. Fruit globose berry, scarlet red, seeds many.
Common in forests and hedges.
Fl. & Fr.: January–May.
Sanskrit: Karambha, cranthila. Telugu: Aridonda, Adonda, Adavinimma.
Propagation: by seeds.

Medicinal uses: The root bark is bitter, stomachic, sedative antihidrotic and useful in fever. The leaves are applied as a poultice to piles, boils and swellings. The plant is useful in fever.

Capsicum annuum L.

Family: Solanaceae.

Annual or slightly perennial, erect bushy herb, glabrous. Leaves simple, ovate-lanceolate. Flowers solitary or in pairs, long-pediceled. Fruit berry 1.5–5cm long, varies in form, red at maturity.
Commonly cultivated all over the state for its fruit.
Sanskrit: Katuirah, Rakta Maricha. Telugu: Mirapakaya.
Trade name: Chillies, Red chillies.

Medicinal uses: Chilli in small doses, is a powerful stimulant and carminative. It is an excellent remedy in atonic and flatulant dyspepsia and dipsomania. Chilli is a good source of vitamin C. Chilli is powerful irritant and rubifacient when applied to the skin. It is used externally for the treatment of rheumatism, lumbago and neuralgia.
Active principle: Capsaicin, dihydrocapsaicin, carotenoids, capsicidine.

Capsicum frutescens L.

Family: Solanaceae.

A perennial shrub, up to 2m tall. Leaves elliptic-ovate. Flowers greenish-white, 2–5 in terminal bunches. Fruit elongate berry, red when ripe, many-seeded, seeds circular, compressed.

Rare in waste lands.
Fl. & Fr.: Round the year.
Sanskrit: Katuvirah. Telugu: Golkonda Mirapa.
Trade name: Golkonda Chillies.
Propagation: by seeds.

Medicinal uses: The fruit is used in back pains and swollen feet. The fruits and bark are reported to be given in jaundice as a remedy. The tincture of the dried fruit is an ingredient in preperations used as an irritant in dandruff. It is also used in liniment and plasters for rheumatism and neuralgia. The chilli vinegar is an excellent stomachic. The leaves form an ingredient in remedies for heatstroke, inflammation and boils.

Caralluma umbellata Haw.

Syn: *Boucerosia umbellata* (Haw.)Wt. & Arn.
Stapelia umbellata (Haw.) Roxb.

Family: Asclepiadaceae.

Erect, succulent leafless herbs, up to 30cm tall. Flowers in terminal umbels. Fruit paired follicles, cylindric, green with small dark spots.

Occasional in hilly regions of Rayalaseema.
Fl.& Fr.: March–August.
Sanskrit: Dugdhika, Uttamphalini. Telugu: Kundeti kommulu.
Propagation: by stems.

Medicinal uses: Stems are used in stomach disorders and abdominal pains.

Cardiospermum halicacabum L.

Family: Sapindaceae.

Annual pubescent climbing tendril bearing herb. Leaves bipinnate, leaflets ovate-lanceolate, deeply incised. Flowers white in axillary racemes with two opposite tendrils.

Fruit inflated capsule, triangled. Seeds black, with heart-shaped hilum.
Common among bushes all over Andhra pradesh.
Fl. & Fr.: July–December.
Sanskrit: Indravalli, Jyotismati, Sakralata.
Trade name: Ballon vine, Heart's pea.

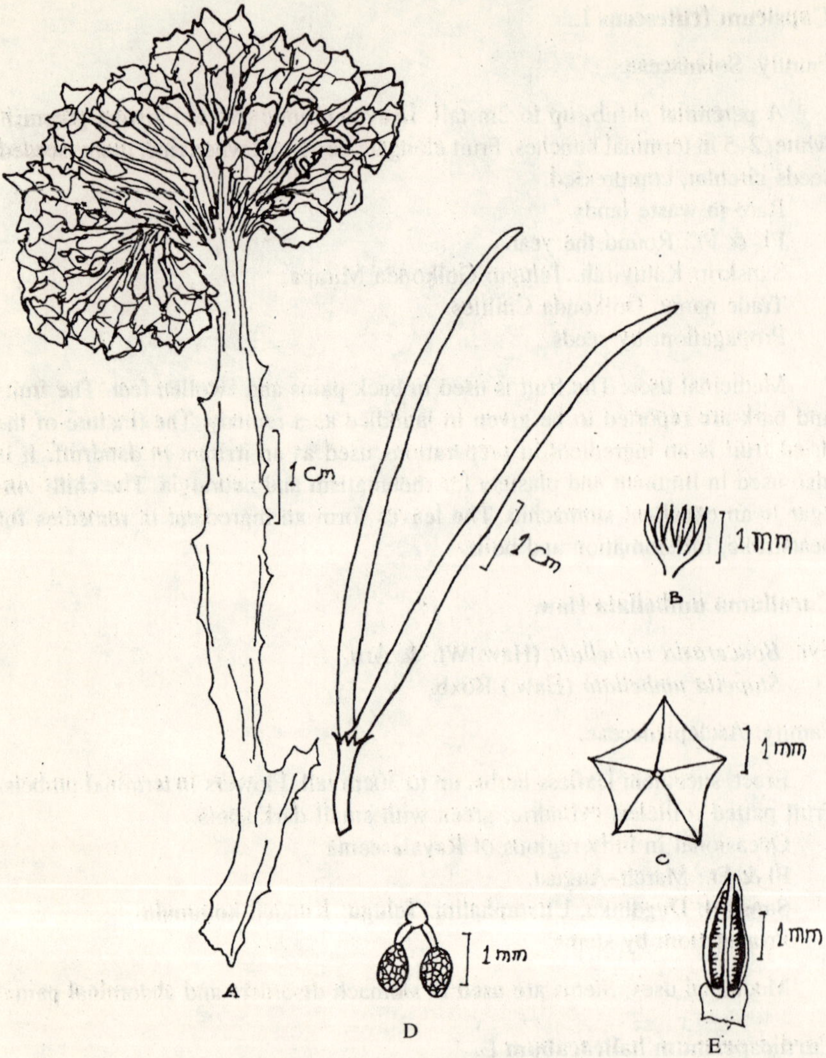

Fig. 29: *Caralluma umbellata* Haw.
A. Twig; B. Calyx split open; C. Corolla split open; D. Pollinia; E. L.S. of ovary.

Medicinal uses: The herb is diuretic, stomachic and rubefacient. It is used in rheumatism, lumbago, nervous diseases, as a demulcent in orchitis and dropsy. The root is mucilaginous and considered effective in didymitis and early stages of hydrocele, and is used in asthma and colic.

Active principle: Saponin.

Careya arborea Roxb.

Family: Barringtoniaceae

Trees, up to 8m tall, young branches gnarled with persistent leaf scars, pubescent. Leaves apically clustered, broadly oblong-ovate. Flowers yellowish white.

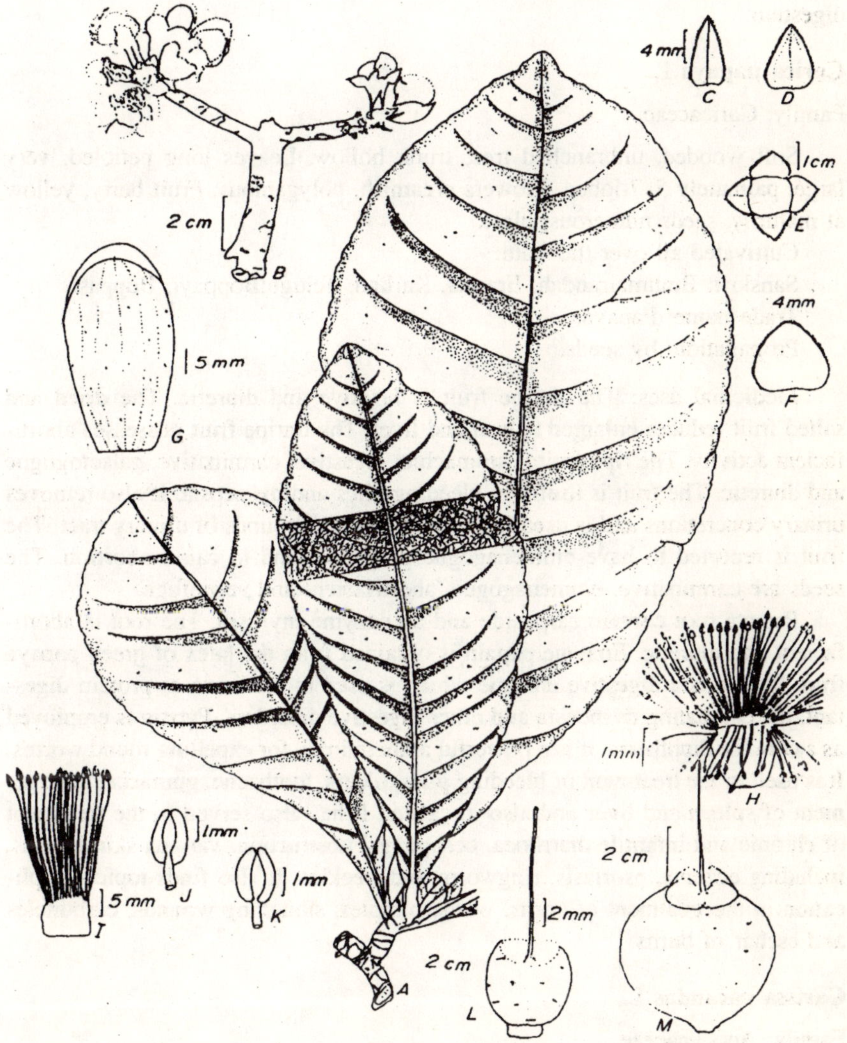

Fig. 30: *Careya arborea* Roxb.

A. Twig; B. Inflorescence; C. & D. Bracts; E. Calyx; F. Sepal; G. Petal; H. Stamen with calyx; I. Stamens; J. & K. Anthers; L. Pistil; M. Fruit.

Medicinal uses: Various parts of the tree (known as padmaka in Ayurvedic system of medicine) are reputed in Indian medicine. The stem bark is used as demulcent in coughs and colds, as antipyretic in eruptive fevers and as anthelmintic. The decoction of the bark is used in diarrhoea and also for washing eyes in eye complaints. The dried calyces (Trade: Vakiembha) are tonic and demulcent. The flowers are used as an aphrodisiac and their infusion is given after child birth. The fruits are aromatic, astringent and demulcent and promote digestion.

Carica papaya L.

Family: Caricaceae.

Soft wooded, unbranched tree, trunk hollow. Leaves long petioled, very large, palmately 5–7-lobed. Flowers creamish, polygamous. Fruit berry, yellow at maturity, seeds numerous, black.

Cultivated all over the state.

Sanskrit: Brahmairandah, Eranda, Karkati. Telugu:Boppayi, Boppasi.

Trade name: Papaya.

Propagation: by seeds.

Medicinal uses: The unripe fruit is laxative and diuretic. The dried and salted fruit reduces enlarged spleen and liver. The unripe fruit possesses abortifacient activity. The ripe fruit is stomachic, digestive, carminative, galactogogue and diuretic. The fruit is useful in bleeding piles and dyspepsia. It also removes urinary concretions and is used in haemoptysis and wounds of urinary tract. The fruit is reported to have emmenagogue properties and to cause abortion. The seeds are carminative, emmenagogue, abortifacient and vermifuge.

Papaya root contain carposide and an enzyme myrosin. The root is abortifacient and diuretic. Enzyme papain is obtained from the latex of green papaya fruit. Papain is a digestive enzyme which is used in medicine as protein digestant, for combating dyspepsia and other digestive disorders. Papain is employed as a cure for diphtheria. It is a powerful anthelmintic, for expelling round worms. It is used in the treatment of bleeding piles, ulcers, toothache, gumache, enlargement of spleen and liver and also in cancer. It has also served in the treatment of chronic and infantile diarrhoea, oesophagal obstruction, various skin lesions, including eczema, psoriasis, ringworm and freckles. It also finds topical application in the treatment of corns, warts, pimples, sloughing wounds, carbuncles and eschar of burns.

Carissa carandas L.

Family: Apocynaceae.

Large evergreen thorny shrub, to 5m tall, spines to 3cm. Leaves elliptic or obovate, coriaceous, glabrous. Flowers white in cymes. Fruit ellipsoid berry, purplish-black when ripe, 4 or more-seeded.

Occasional in forests in Srikakulam, Visakhapatnam, Krishna and Cuddapah districts.

Fl. & Fr.: January–June.

Sanskrit: Kanachuka, Karamarda. Telugu: Wakkayalu.

Trade name: Karaunda.

Medicinal uses: The ripe fruits are antiscorbutic, cooling and useful in bilious complaints. All parts of the plant are reputed in indigenous medicine. The roots are credited with bitter stomachic and anthelmintic properties.

Carissa spinarum L.

Family: Apocynaceae.

Armed shrub, to 3m tall, spines to 2cm long, sometimes forked. Leaves elliptic-ovate. Flowers white, sometimes tinged pink, scented, in few-flowered terminal cymes. Fruit globose berry, purplish black when ripe.

In scrub jungles in Rayalaseema and southern Telangana.

Fl. & Fr.: September–December.

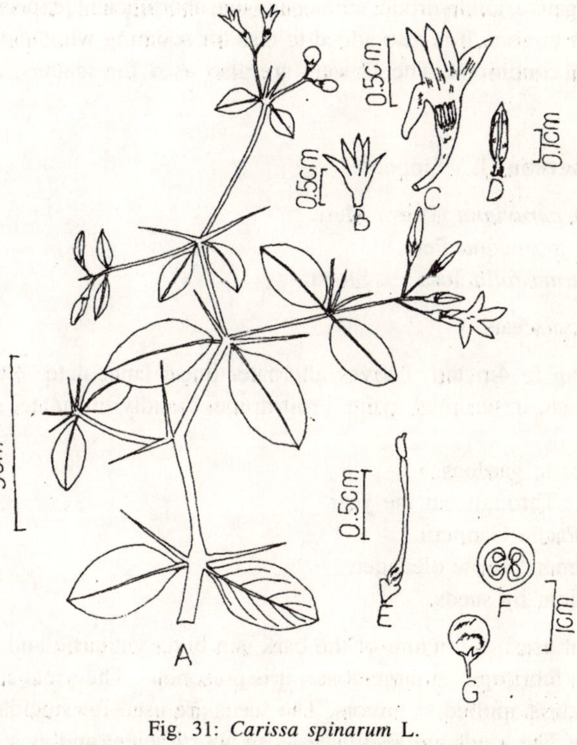

Fig. 31: *Carissa spinarum* L.

A. Twig; B. Calyx split open; C. Corolla split open; D. Stamens; E. Pistil; F. T.S. of ovary; G. Fruit.

Sanskrit: Bahudala, Kantaki. Telugu: Kalivi.

Medicinal uses: The roots have been atributed several uses in indigenous medicine. They possess purgative property. Leaves are used in intermittent fevers and general pains.

Carthamus tinctorius L.

Family: Asteraceae (Compositae)

Annual erect herb, up to 1m tall. Leaves alternate, spiny. Flowers orange yellow in terminal heads. Seeds white, shiny.

Cultivated in black soils in dry lands.

Sanskrit: Kusumbha. Telugu: Kusuma.

Trade name: Safflower.

Propagation: by seeds.

Medicinal uses: The plant is laxative, stimulates menstrual flow and induces perspiration. Paste made from flower is applied to boils. Tea made from flower is used to treat fever and skin eruptions. In folk medicine flowers are used as stimulant, pargative, antihydrotic, emmenagogue, abortifacient, expectorant, pneumonic and for tumors. It is also added to teas for soothing which soothe coughs and bronchial conditions. The flowers are also used for scabies, arthritis and chest pains.

Cascabela thevetia (L.) Lippold

Syn: *Thevetia peruviana* (Pers.) Merr.
Cerbera peruviana Pers.
Thevetia nerifolia Juss. ex Steud.

Family: Apocynaceae

Shrubs, up to 4m tall. Leaves alternate, linear-lanceolate. Flowers large, bright yellowish, in terminal cyme. Fruit drupe, broadly turbinate, seeds a few, flattened.

Cultivated in gardens.

Fl. & Fr.: Through out the year.

Telugu: Pacha Ganneru.

Trade name: Yellow oleander.

Propagation: by seeds.

Medicinal uses: A tincture of the bark is a bitter cathartic and emetic, it is also used as a febrifuge. In large doses, it is poisonous. The roots are made into a plaster which is applied to tumors. The seeds are used for suicidal and homicidal purpose. The seeds are also used as an abortifacient and as a purgative in rheumatism and dropsy.

Cassia absus L.

Family: Leguminosae (Fabaceae)

Erect annual herbs, up to 60cm tall, stem and leaves clothed with grey bristly viscous hairs. Leaves long petioled, pinnate, leaflets 2 pairs, ovate-oblong or ovate-elliptic, glabrous above, pubescent below. Flowers pure yellow or tinged red, in terminal or leaf-oppossed racemes. Fruit pod, flat, covered with stiff glandular hairs, 5–7-seeded, seeds black, ovoid.

In waste lands and in open forests.

Fl. & Fr.: August–February.

Medicinal uses: The seeds are slightly mucilaginous and bitter, astringent, stimulant and diuretic. They are employed as a cathartic in habitual constipation. They are also used for ringworm, venereal ulcers and other diseases of skin, for pains, headache, bronchitis; leucoderma, asthma, haemorrhoids and as anthelmintic. The decoction of seeds is used in eye diseases.

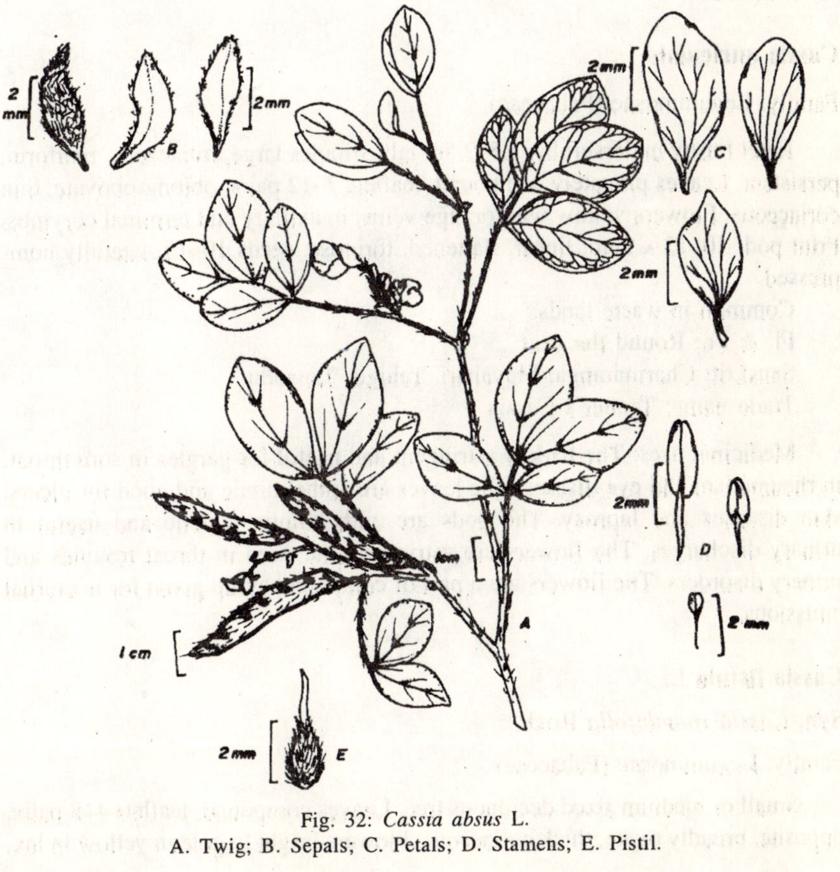

Fig. 32: *Cassia absus* L.
A. Twig; B. Sepals; C. Petals; D. Stamens; E. Pistil.

Cassia alata L.

Family: Leguminosae (Fabaceae).

Shrubs, up to 2m tall, branchlets glabrescent. Leaves pinnately compound, leaflets 8–12 pairs, oblong, inequilateral, thin coriaceous, glabrous, rachis narrowly winged. Flowers yellow in terminal dense racemes, bract spathaceous. Fruit pod, oblong, thin, straight, seeds 50 or more.

Cultivated, sometimes run wild.

Trade name: Ringworm cassia.

Medicinal uses: The leaves are employed as a purgative in the same manner as senna. They are reported to have expectorant, diuretic, sudorific properties. The fresh leaf or its juice or the decoction is widely employed as a remedy in other skin diseases like herpes, blotch eczema and mycosis and for treating syphilis and gonorrhoea. The dried leaves are reported to be used against leprosy. A decoction of the leaves and flowers is useful in bronchitis, asthma and in stomatitis.

Cassia auriculata L.

Family: Leguminosae (Fabaceae)

Erect bushy undershrubs, 1.5–2.5m tall, stipules large, foliaceous, reniform, persistent. Leaves pinnately compound, leaflets 7–12 pairs, oblong-obovate, thin coriaceous. Flowers yellow with orange veins, in axillary and terminal corymbs. Fruit pod, 10–12 x 2cm, linear, flattened, torulose, seeds 10–15, laterally compressed.

Common in waste lands.

Fl. & Fr.: Round the year.

Sanskrit: Charmaranga, Mayahari. Telugu: Tangedu.

Trade name: Tanner's Cassia.

Medicinal uses: The bark is astringent and useful for gargles in sore throat, in rheumatism and eye disases. The leaves are anthelmintic and good for ulcers, skin diseases and leprosy. The pods are anthelmintic, emetic and useful in urinary discharges. The flowers are astringent and used in throat troubles and urinary disorders. The flowers are a part of compound syrup given for nocturnal emissions.

Cassia fistula L.

Syn: *Cassia rhombifolia* Roxb.

Family: Leguminosae (Fabaceae)

Small or medium sized deciduous tree. Leaves compound, leaflets 4–8 pairs, opposite, broadly ovate, thick coriaceous. Flowers bright to golden yellow in lax,

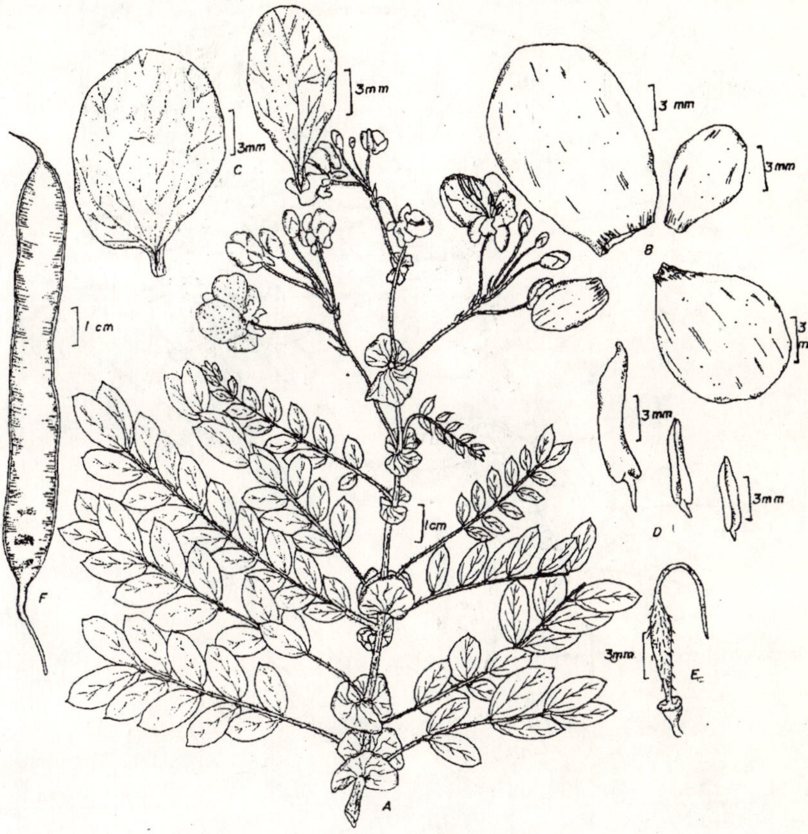

Fig. 33: *Cassia auriculata* L.
A. Twig; B. Sepals; C. Petals; D. Stamens; E. Pistil; F. Pod.

pendent, branched raceme. Fruit 40–50cm long pod, cylindric, dark blackish-brown.

Common in deciduous forests all over the state.

Fl. March–May Fr.: May–December.

Sanskrit: Aragvadha, Suvarnaka. Telugu: Rela.

Trade name: Rela.

Propagation: by seeds.

Medicinal uses: The pulp from the fruits, called Cassia pulp, is a well known laxative. The bark possesses tonic and antidysenteric properties. It is also used in skin complaints. The wood is given in dysentery. The powder or docoction of the bark is administered in leprosy, jaundice, syphilis and heart diseases. The root and its bark possess astringent, tonic, febrifugal and purgative properties.

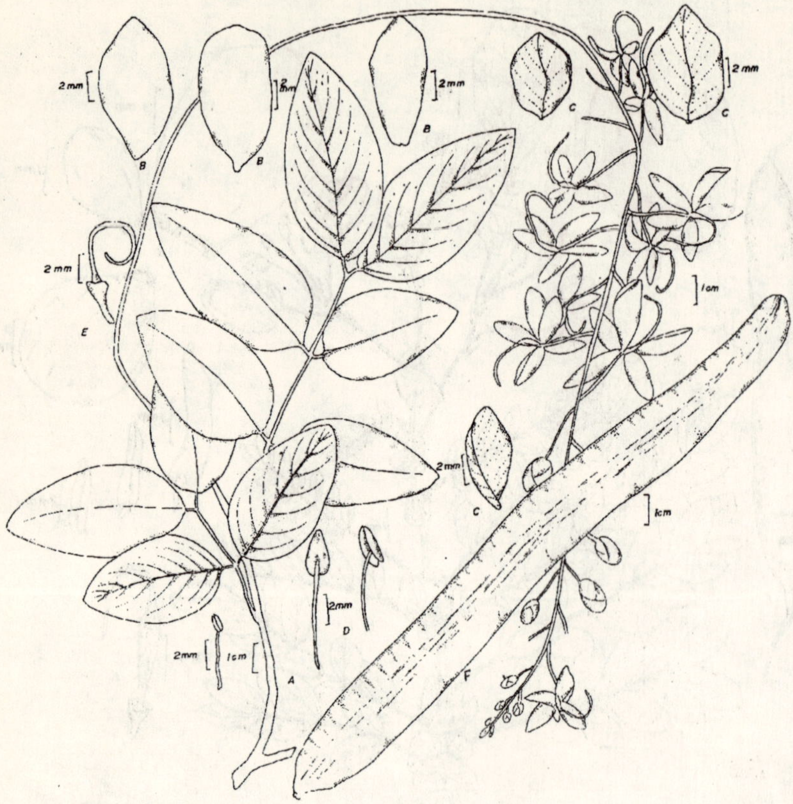

Fig. 34: *Cassia fistula* L.
A. Twig; B. Sepals; C. Petals; D. Stamens; E. Pistil; F. Pod.

Cassia occidentalis L.

Family: Leguminosae (Fabaceae)

Erect, glabrous undershrus, up to 3m tall, branchlets glabrous. Leaves 10–15cm long, paripinnate, leaflets 3–5 pairs, elliptic-ovate, membranous, glabrous above, thinly hairy beneath. Flowers bright reddish yellow in axillary and terminal corymbose racemes. Fruit pod, brown with yellow margins, seeds 20–25, greenish brown.

Common on road sides and wastelands all over the state.

Fl. & Fr.: Through out the year.

Sanskrit: Kasam arda, Kanaka. Telugu: Kasinda.

Medicinal uses: All parts of the plant possess purgative, tonic, febrifugal, expectorant and diuretic properties. The plant is used to cure sore eyes, haematuria, rheumatism, typhoid, asthma and disorders of haemoglobin. A decoction of the

Fig. 35: *Cassia occidentalis* L.
A. Twig; B. Sepals; C. Petals; D. Stamens; E. Pistil.

plant is used in hysteria, in dysentery and other stomach troubles, and also as an appliction to sores, itch and inflammation of rectum. The plant is employed in dropsy and as a vermifuge. The herb forms an ingredient of the patented indigenous herbal drug "Liv-52", which shows marked effect in the early cases of hepatic cirrhosis having ciatorrhoea.

Cassia senna L.

Syn: *Cassia angustifolia* Vahl
Cassia obovata Baker

Family: Leguminosae.

A small perennial undershrub, with pale-yellowish brown bark. Leaves alternate, compound, leaflets 4–7 pairs, opposite, linear-lanceolate, glabrous above, appressed pubescent below. Flowers greenish-yellow in terminal and axillary racemes. Fruit flat legume, greenish brown to dark brown, seeds dark brown.

Cultivated for its cathartic leaves and pods.

Fl. & Fr.: July–September.

Fig. 36: *Cassia senna* L.
A. Twig; B. Sepals; C. Petals; D. Staminodes; E. Stamens; F. Pistil.

Sanskrit: Markundi. Telugu: Sunamukhi, Cinemanaku.

Trade name: Indian Senna.

Propagation: by seeds.

Medicinal uses: Leaves and pods contain sennosides. It is used as a laxative or purgative since ancient times in unani system. It is reported to be a safe and effective drug against habitual constipation, abdominal disorders, leprosy, skin diseases, leucoderma, jaundice, dyspepsia, cough, bronchitis, typhoid fever, anaemia and tumors.

Cassia sophera L.

Family: Leguminosae (Fabaceae).

Diffuse undershrubs, up to 3m tall. Leaves 10–15cm long, leaflets 6–10 pairs, oblong-lanceolate, glaucous. Flowers yellow in terminal and axillary racemes. Fruit pod, linear-oblong, 7–12 x 0.6–0.8cm, septate, seed ovoid, flat, dark brown.

Fig. 37: *Cassia sophera* L.
A. Twig; B. Sepals; C. Petals; D. Stamens; E. Pistil.

Common in all plains districts, by road sides and on waste lands.
Fl. & Fr.: July–January.
Sanskrit: Kasamanda, Suvarnamayaheri. Telugu: Kondakasinda, Pyditangedu.
Trade name: Senna sophera, Kasa marda.

Medicinal uses: The plant exhibited anti cancer activity. It is reported to be extensively used in Homeopathy. A decoction of the plant is antidiuretic. The leaves possess purgative properties. Their juice made into a paste with sandal-wood or mixed with lime juice is considered specific for ring worm. Externally is is used for washing syphilitic sores. The juice in honey or infusion or decoction is administered as an expectorant for cough, cold, bronchitis, asthma and hiccups.

Cassia tora L.

Family: Leguminosae. (Fabaceae)

Annual herbs, up to 20cm tall, crisped hairy in young branches. Leaves 6–8cm long, pinnately compound, leaflets 3–5 pairs, sessile, obovate-oblong, thin

Fig. 38: *Cassia tora* L.
A. Twig; B. Sepals; C. Petals; D. Stamens; E. Pistil.

coriaceous. Flowers golden yellow or with reddish tinge, solitary or geminate, axillary. Fruit pod, linear, puberulous, 10–15 x 0.3cm, seeds about 20, faintly tetragonal.

Common on waysides and waste places and fallow lands.

Fl. & Fr.: September–December.

Sanskrit: Chakramardah, Prapunnatah. Telugu: Tantmu, Tagirisa.

Trade name: Foetid Cassia.

Medicinal uses: Chakramardha tailamu, a compound ayurvedic oil of this herb, is beneficial in eczema, ringworm and other skin diseases. The herb possesses purgative and anthelmintic properties. A paste of pulverzed seeds with lime juice is used for ring worm and other skin-diseases. Seeds are also used in eye-diseases, liver complaints and earache.

Casuarina equisetifolia L.

Family: Casuarinaceae.

Tall, leafless tree, up to 15m tall, branchlets arising from the axils of scales. Leaves scaly, alternating with ribs of the upper node. Flowers unisexual. Male flowers in terminal catkins, female in globose-ovoid cone like heads. Fruit nutlets, compressed, winged.

Widely planted along the coast.

Fl. & Fr.: February–August.

Telugu: Sarugudu.

Trade name: Beef wood.

Propagation: by seeds.

Medicinal uses: The bark is a tonic and astringent. It is useful in diarrhoea and dysentery. Its powder is used for the treatment of acne. A lotion of it is reported to be efficaceous as gargle for sore throat, in beri-beri, in coughs, ulcers, constipation and stomachache. The decoction of needles is given in colic and used as a lotion for swellings.

Catharanthus pusillus (Murr.) G.Don

Syn: *Vinca pusilla* Murr.

Lochnera pusilla (Murr.) K.Schum.

Family: Apocynaceae.

Annual erect herb, to 30 cm tall, branchlets terete below, quadragular above. Leaves lanceolate, glabrous. Flowers white in terminal or axillary 5–7-flowered corymbs. Fruit capsule, seeds black.

Occasional weed in cultivated fields in all districts.

Fl. & Fr.: July–December.

Sanskrit: Sangki. Telugu: Errimirapa, Kondamirapa.

Propagation: by seeds.

Medicinal uses: The plant possesses oncolytic (tumor resolving) properties. A decoction of the dried plant boiled in oil is used in the treatment of lumbago. Whole plant used in paralysis, epilepsy and ulcers.

Catharanthus roseus (L.) G.Don

Syn: *Vinca rosea* L.

Lochnera rosea (L.) Reichnb.

Family: Apocynaceae.

A perennial undershrub, up to 50cm tall. Leaves elliptic obovate to oblong, opposite decussate. Flowers pink or white, solitary or paired, axillary. Fruit a pair of follicles, many-seeded.

Cultivated all over the state mainly as an ornamental, some times for medicine.

Fl. & Fr.: Round the year.

Sanskrit: Nityakalyani, Shanthapushpa. Telugu: Billa ganneru.

Trade name: Madagascar periwinkle.

Propagation: It is propagated by seeds. The plants are first grown in nursery. In irrigated lands 1–1.5 kg of seeds is sown in nurseries in May–June. The seeds germinate in 10 days. In the early days the seed beds should be covered with hay or phoenix leaves. If it is irrigated daily it can be transplanted in 45–60 days in one hectare of land.

The soil is ploughed and 5–10 tonnes of farm yard manure, 200 kg of superphosphate 50 kg of potash is added to the soil. In irrigated lands the distance between rows is 30–45 cm, between plants is 30 cm (74,000–1,11,000 per hectare). In dry lands 4–5 kgs of seeds are soaked in water during night, mixed with 8–10 kgs of sand and then sown. In irrigated lands the land should be irrigated once in 10–15 days. 20–25 days after sowing weeds are removed. Thereafter depending on the situation the weeds should be removed 2–3 times. In dry lands if there are more plants in a clump, they should be taken out and planted elsewhere. There is a misconception that if chemical fertilizers are used the medicinal properties will get affected, it has been proved in field experiments that this is a wrong concept. 110–120 kg of urea per hectare in rainfed crop and 220 kg of urea per hectare in irrigated crop is added in 3 split doses.

First harvest of leaves can be made 6 months after sowing while subsequent harvest can be made for every 3 months. Generally crop of catharanthus is harvested between 10–12 months. The plants are cut to the base, irrigated, ploughed and roots are collected. The roots are cleaned and dried. Only after finalising the sale, the roots should be removed. Yield in dry lands per hectare is 1000 kg of leaves and 500 kg of roots while irrigated lands yield 3000 kg of leaves and 800 kg of roots.

Medicinal uses: The roots, which form the drug, contain more than 80 alkaloids of which Indole–Indoline dimers (about 20) are pharmacologically very important. The principal alkaloids in this group are vinblastine, vincristine, vinleurosine and vinrosidine. The other important alkaloids of the roots are ajamalacine, tetrahydroalstonine, serpentine and lochnirine.

Vinca alkaloids received great attention due to their anticancer potential. Vincristine and Vinblastine are used as anticancer drugs especially for curing leukemia (blood cancer).

The whole plant is hypoglycaemic.

Catunaregam spinosa (Thunb.) Tirv.

Syn: *Gardenia spinosa* Thunb.
Randia dumetorum (Retz.) Poir.
Gardenia dumetorum Retz.
Randia brandisii Gamble

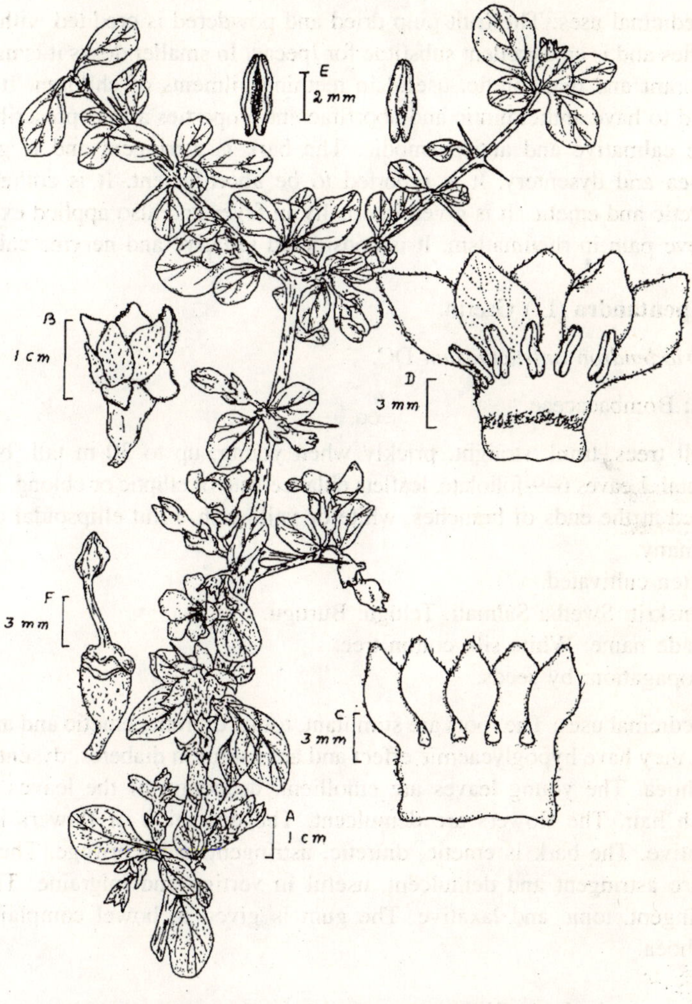

Fig. 39: *Catunaregum spinosa* (Thunb.) Tirveng.
A. Twig; B. Flower; C. Calyx; D. Corolla; E. Stamen-ventral view and dorsal view; F. Pistil.

Family: Rubiaceae

A small armed shrub. Leaves elliptic lanceolate or obovate, sparsely pubescent above, densely pubescent below. Flowers in axillary corymbs, greenish white. Fruit globose berry.

Common in scrub jungles in most districts.

Fl. & Fr.: February–June.

Sanskrit: Madana, Shalya. Telugu: Manga.

Trade name: Emetic nut.

Propagation: by seeds.

Medicinal uses: The fruit pulp dried and powdered is credited with emetic properties and is an excellent substitue for Ipecac. In smaller doses it is nauseant, expectorant and diaphoretic, useful in teething ailments of children. It is also reported to have anthelmintic and abortifacient properties and is prescribed as a nervine calmative and antispasmodic. The bark is astringent and is given in diarrhoea and dysentery. It is reported to be abortifacient. It is anthelmintic, antipyretic and emetic. It is given internally in fever and also applied externally to relieve pain in rheumatism. It is considered sedative and nervine calmative.

Ceiba pentandra (L.) Gaertn.

Syn: *Eriodendron anfructuosum* DC.

Family: Bombacaceae.

Tall trees, trunk straight, prickly when young, up to 30 m tall, branches horizontal. Leaves 6–9-foliolate, leaflets oblanceolate or elliptic or oblong. Flowers clustered at the ends of branches, white or yellowish. Fruit ellipsoidal capsule, seeds many.

Often cultivated.

Sanskrit: Swetha Salmali. Telugu: Burugu.

Trade name: White silk cotton tree.

Propagation: by seeds.

Medicinal uses: The roots are stimulant, tonic, diuretic, emetic and antispasmodic, they have hypoglycaemic effect and are useful in diabetes, dysentery and gonorrhoea. The young leaves are emollient, decoction of the leaves is used to wash hair. The flowers are demulcent. The decoction of flowers is taken as laxative. The bark is emetic, diuretic, astringent and febrifuge. The unripe pods are astringent and demulcent, useful in vertigo and migraine. The gum is astringent, tonic and laxative. The gum is given in bowel complaints and gonorrhoea.

Celastrus paniculatus Willd.

Family: Celastraceae

Climbing shrubs, stem dark pink. Leaves alternate, ovate or obovate. Flowers pale yellow in terminal paniculate cymes. Fruit globose capsule, orange when ripe.

Common in forests.

Sanskrit: Jyothismati. Telugu: Maner tiga, Dati chettu.

Trade name: Climbing staff plant, Intellect tree, Black oil plant.

Medicinal uses: The seeds constitue the drug. They are bitter and have an unpleasent odour. They possess emetic, diaphoretic, febrifugal and nervine properties and are used for sharpening the memory. They are also used to cure sores, ulcers, rheumatism and gout.

The seed oil is used in scabies, body and rheumatic pains, wounds, eczema, beriberi and paralysis. As chief ingredient of drugs the oil showed good results in treating mental depression and gave excellent results in hysteria without any side reaction.

Active principle: The seeds are reported to contain the alkaloids celastrine and paniculatin.

Celosia argentea L.

Family: Amaranthaceae.

Annual erect herb, up to 2m tall. Lower leaves elliptic-ovate, upper ones linear-lanceolate. Flowers white in terminal and axillary dense spikes. Fruit utricle, circumscissile.

Common weed in post harvested dry fields and waste lands.

Fl. & Fr.: August–December.

Sanskrit: Vitunna, Telugu: Gurugu.

Trade name: Silver-spiked cocks comb, Quail grass.

Propagation: by seeds.

Medicinal uses: The paste of leaves is applied on scorpion sting. The flowers are useful in blood dysentery, spitting of blood and menorrahagia. The seeds are considered aphrodisiac and efficacious in diarrhoea. They are reported to be useful in blood diseases, tumors, mouth sores and in eye diseases.

Centella asiatica (L.) Urban.

Syn: *Hydrocotyle asiatica* L.

Family: Umbelliferae (Apiaceae).

A perennial prostrate herb, creeping with long stolons, rooting at nodes. Leaves simple, orbicular-reniform, puberulous. Flowers pink or red in axillary fasciculate umbels, central flowers sessile, laterals pedicellate. Fruit ovoid, mericarp 7–9-ribbed, prominently reticulate.

Common in wet places in most districts.

Fl. & Fr.: Through out the year.

Sanskrit: Mandukaparni. Telugu: Saraswathi aku.

Trade name: Indian penny wort.

Propagation: Rooted stem cuttings and by seeds.

Medicinal uses: Whole plant used in anxiety, neurosis, memory enhancer, general debility, jaundice, leprosy and skin diseases. The plant is used as an antidote to cholera. A cold poultice of the fresh herb is used as an external application in rheumatism, elephatiasis and hydrocele. For treating leprosy and other skin diseases it is given as an ointment or dusting powder. Internally it has been valued as a tonic and is used in bronchitis, asthma, gastric catarrh, leucorrhoea, kidney troubles, urethoritis and dropsy. The plant is considered as aphrodisiac and used in venereal diseases.

Contain Asiaticosides is used in skin diseases and psychotropic. Asiaticoside is also useful in the treatment of leprosy and certain types of tuberculosis.

Cerbera odollam Gaertn.

Family: Apocynaceae.

Moderate sized tree or large shrub with an acrid milky poisonous juice. Leaves alternate, lanceolate or oblanceolate. Flowers large, white with an yellow throat, odorous, in terminal peduncled paniculate cymes. Fruit drupe, subglobose, green, seed solitary.

East coast in Salt swamps and backwaters. Also cultivated as an ornamental.

Fl. & Fr.: June–January.

Sanskrit: Auddalakah, Svanamarh.

Telugu: Adava Arali.

Trade name: Dog-Bane, Odallam Tree.

Propagation: by seeds

Medicinal uses: All parts of the plant contain cardenolide type cardioactive glycosides. The kernels are highly poisonous and have been commonly used for suicidal and homicidal purposes.

The bark and the latex of the leaves are emetic and cathartic. The leaves are cooked and eaten for their mild laxative effect. The latex contains a poisonous glycoside thevetin.

Cereus pterogonus Lemaire

Family: Cactaceae

Armed leafless cactus, up to 6m tall, stems columnar, 3–9-winged or angled. Flowers white, lateral, sessile.

Common in field hedges.

Fl: March–July. Fruits seldom seen.

Telugu: Bontha Jamudu.

Propagation: by stem cuttings.

Medicinal uses: Fresh young shoots are used as cardiac stimulant, in dropsy and cardiac affections.

Chlorophytum arundianceum Baker

Family: Liliaceae.

Rhizomatous herbs with tuberous roots. Leaves suberect, lanceolate. Flowers white in dense flowered racemes. Fruit capsule, 2-lobed at tip and base, cells 3–4-seeded, seeds flat and black.

Cultivated.

Fl. & Fr.: June–September.

Sanskrit: Sweta musli.

Trade name: Safed musli.

Propagation: by seeds and roots.

Medicinal uses: The tubers constitute the drug, commercially known as safed Musli. The drug is considered a valuable nervine and general tonic for strength and vigour. The fried powder of the root is chewed in aphthae of mouth and throat. The decoction of the root with turmeric is given in rheumatism.

Chlorophytum tuberosum (Roxb.) Baker

Syn: *Anthericum tuberosum* Roxb.

Family: Liliaceae.

Herbs with cylindric root-fibres and ellipsoid tubers hanging from them. Leaves 6–12, membranous, sessile, usually falcate, recurved, margin undulate. Scape terete. Flowers white in simple or shortly branched racemes. Fruit capsule, oblong, triquetrous strongly nerved. Seeds subquadrate, irregularly pitted.

In all districts, a weed in cultivated fields and wastelands.

Fl. & Fr.: June–November.

Sanskrit: Musli.

Trade name: Safed Musli.

Propagation: by tubers.

Medicinal uses: The tuberous roots are reported to be used as a non-hormonal restorative tonic for men and women. It is also used as a general purpose tonic against fatigue, weakness and general debility. The tubers are reported to be used to treat leucorrhoea.

Cissampelos pareira L.

Family: Menispermaceae

Slender climbers with hairy branches. Leaves reniform, orbicular or cordate. Flowers unisexual, greenish yellow, minute, male in panicles, female in pendulous racemes in axils of conspicuous bracts. Fruit ovoid drupe, scarlet.

Common in most districts.

Fl. & Fr.: April–October.

Sanskrit: Patha, Vriddhakarnika. Telugu: Adavibankateega.

Trade name: Velvet-leaf pareira, False pareira brava or root.

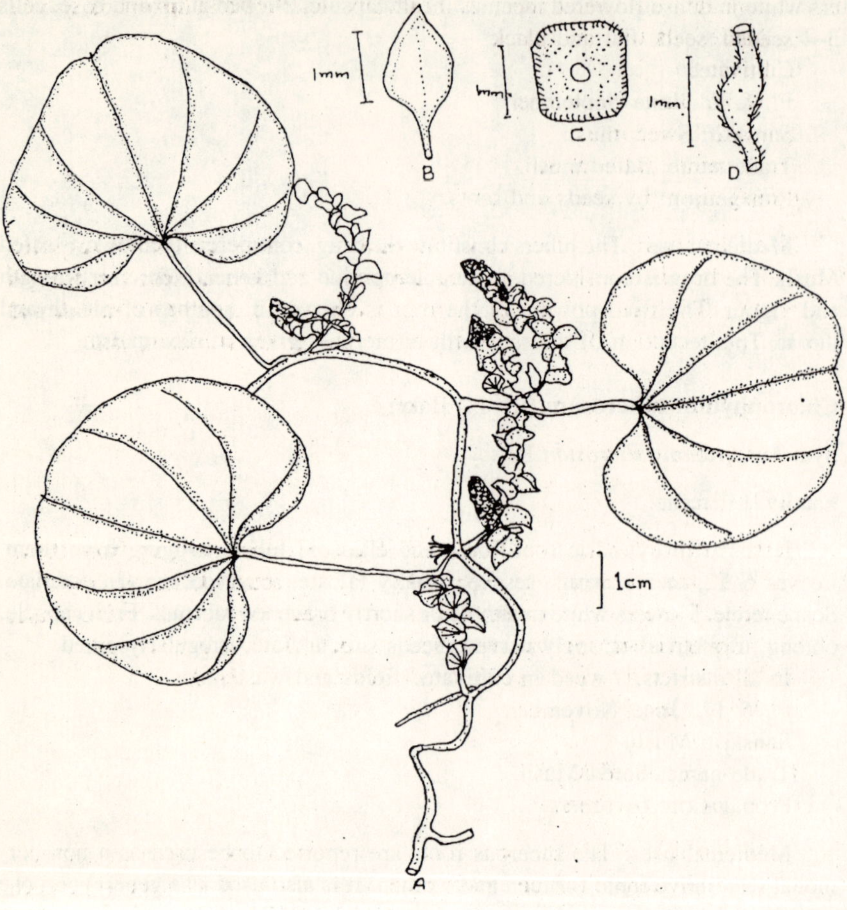

Fig. 40: *Cissampelos pareira*
A. Female Twig; B. Sepal; C. Petal; D. Pistil.

Medicinal uses: The dried roots form the drug. The roots possess astringent, mild tonic, diuretic, stomachic, antilithic, analgesic, anti-pyretic and emmenagogue properties. They are frequently prescribed for treating cough, dyspepsia, diarrhoea, dysentery, piles, dropsy, urino-genital troubles such as prolapses uteri, cystitis, haemorrhage and menorrhagia and calcular nephritis.

Chemistry: The roots and leaves contain several alkaloids.

Cissus pallida (Wt. & Arn.) Planch.

Syn: *Vitis pallida* Wt. & Arn.

Family: Vitaceae.

Rambling shrubs, stems reddish, striate, tend... leaf-opposed. Leaves simple, orbicular, thin coriaceous. Flowers pale yellow in compound umbellate cymes. Fruit globose berry.

Common in forests.

Fl. & Fr.: January–December.

Telugu: Nallateega.

Propagation: Stem cuttings.

Medicinal uses: Bruised roots applied to rheumatic swellings.

Cissus quadrangularis L.

Syn: *Vitis quadrangularis* (L.) Wight & Arn.

Family: Vitaceae.

Rambling shrubs, stem fleshy, 4-angular, glabrous, winged or margined, contracted at nodes. Leaves simple, ovate or reniform, thick coriaceous, tendril leaf-opposed, highly coiled. Flowers greenish-red, in leaf-opposed, short peducled, umbellate cymes. Berry globose, reddish brown, seed solitary, obovate, smooth.

Common in scrub jungles, roadsides and wastelands in all districts.

Fl. & Fr.; June–December.

Sanskrit: Asthisamhari, Asthisrinkala, Asthi sandhana. Telugu: Nalleru.

Trade name: Edible stem vine.

Propagation: Stem cuttings with 3–4 nodes.

Medicinal uses: It is known as 'Hadjor' in trade which means" Joining bones". Powdered stem and root paste is used in bone fractures. Stem and whole plant are useful in asthma, bowel complaints, epitaxis, scurvy and irregular menstruation. The stem is useful in piles and its juice is beneficial as alterative in scurvy and in diseases of the ear and in nose bleeding.

Citrullus colocynthis (L.) Schrad.

Syn: *Colocynthis vulgaris* Schrad.

Family: Cucurbitaceae.

Trailing scabrid herbs, tendrils slender, simple or 2–3-fid. Leaves ovate or narrowly triangular, deeply 3–7-lobed, lobes pinnatifid, densely villous. Flowers unisexual, solitary. Fruit globose, striped green and white when young, yellow when ripe, seeds yellowish-brown.

Common in sandy or stony lands in Telangana and Rayalaseema.

Fl. & Fr.: June–December.

Sanskrit: Chitraphala, Indravarooni. Telugu: Etipuccha, Verripuccha.

Trade name: Colocynth, Bitter apple.

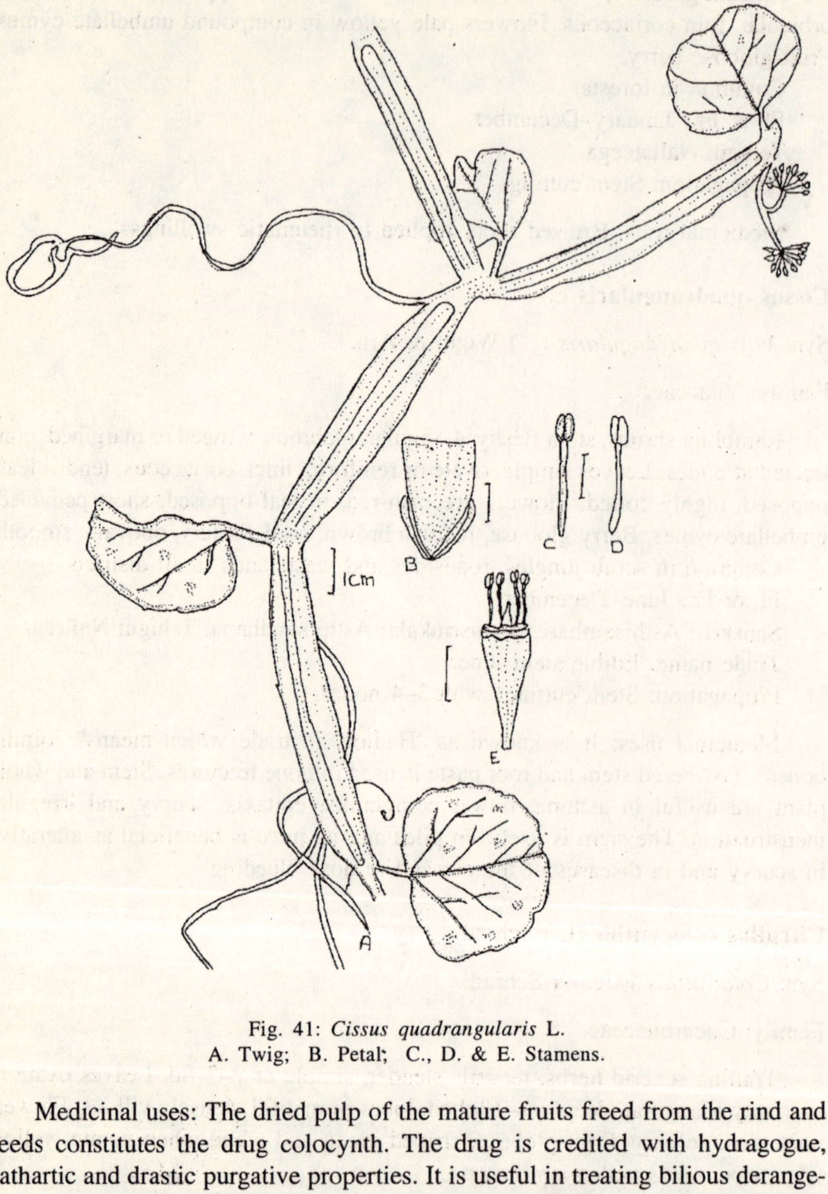

Fig. 41: *Cissus quadrangularis* L.
A. Twig; B. Petal; C., D. & E. Stamens.

Medicinal uses: The dried pulp of the mature fruits freed from the rind and seeds constitutes the drug colocynth. The drug is credited with hydragogue, cathartic and drastic purgative properties. It is useful in treating bilious derangements of chronic constipation and dropsy, fevers and cases requiring purgativeness. It is considered emmenagogue and a vermifuge. The pulp is also used for varicose veins and for piles. It is used as a gargle and mouth wash, a counter-irritant in chest cold. It is diuretic and in small doses expectorant.

Active principles: Cucurbitacins, chlorogenic acid.

Citrus aurantifolia (Christm.) Swingle

Family: Rutaceae.

Armed shrub or small tree, 2–3 m tall. Leaves elliptic-ovate, petiole winged. Flowers white or pink in lax axillary racemes. Fruit round to oval, greenish yellow when ripe, rind very thin.

Cultivated extensively for its fruit.

Sanskrit: Nimbuka. Bribat Jambhira.

Telugu: Nimma.

Trade name: Acid lime, Sour lime, Bitter lime.

Propagation: by seeds.

Medicinal uses: Lime is rich in vitamins, minerals and alkaline salts. Fruit juice is a good source of Vitamin C. It is considered aperient, astringent, antipyretic and sedative. The fruits are regarded as antiseptic, styptic and sudorific.

Citrus aurantium L.

Family: Rutaceae.

Armed trees, up to 8 m tall with a rounded crown and smooth greyish-brown bark. Leaves ovate, petioles are broadly winged, tapering to a wingless base. Flowers white in axillary cymes, sweet scented. Fruit subglobose, loose skinned, orange when ripe, with hollow core when ripe.

Rarely cultivated.

Telugu: Mallikanaragi.

Trade name: Bitter orange, Sour orange or seville orange.

Propagation: by seeds, cuttings, budding and grafting.

Medicinal uses: Preparations of Bitter orange flower and flower oil are used as a preventive measure for gastric and nervous complaints, gout, sore throat, as a sedative, for nervous tension and sleeplessness. The leaves are prescribed in the treatment of swollen legs, in bronchitis and as a tonic. The leaves, peel and flowers are used as a stomachic and antiscorbutic. Sour orange peel is an ingredient in the preperation of tinctures of cinchona and gentian. A decoction of peel is used as an emmenagogue. The dried rind is valuable in atonic dyspepsia and general debility. The fresh rind is rubbed on the face for acne and also on parts affected by eczema. The oil from the rind is internally stomachic and externally a stimulant, lininent, useful in gout and rheumatism.

Citrus medica L.

Family: Rutaceae.

Shrubs or small trees, young shoots glabrous, purple. Leaves simple, petiole winged, oblong. Flowers white, usually tinged with red in raceme. Fruit large, oblong or obovoid, rind usually warted, thick, pulp scanty.

Rare in Visakhapatnam and Prakasam districts.
Sanskrit: Matulunga. Telugu: Madiphalam.
Trade name: Citron.
Propagation: Air layering.

Medicinal uses: Juice of fruits is refrigerant and astringent. Preserved rind is used in dysentery. The fruits are stomachic and have potent anti-scorbutic activity. Fruit decoction is analgesic, it is effective in vomitting. The fruit and seeds are a cardiac tonic and useful in palpitation.

Cleome viscosa L.

Family: Capparaceae.

Annual erect, branched, viscid pubescent herb. Leaves 3–5-foliolate, leaflets obovate or elliptic. Flowers yellow in axillary or terminal racemes. Fruit terete capsule, seeds numerous, dark brown, sub-globose.
Common weed of waste lands in all districts.
Fl. & Fr.: June–December.
Sanskrit: Adityabhakta, Arkakanta. Telugu: Kukkavaminta.
Trade name: Wild mustard, Dog mustard.
Propagation: by seeds.

Medicinal uses: The leaves are rubifacient, vesicant and sudorific. The juice of the leaves mixed with ghee is used in the treatment of inflammations of the middle ear. The leaves are also used in external applications for wounds and ulcers. The seeds are reported to have rubifacient, vesicant and anthelmintic properties. A poultice made from seeds is efficaceous as a counterirritant in chronic painful joints.

Clerodendrum phlomides L.

Family: Verbenaceae.

A shrub, up to 3 m tall, branchlets pubescent. Leaves deltoid-ovoid. Flowers cream in axillary or terminal panicles. Fruit obovoid drupe.
Common in hedges near villages in most districts.
Fl. & Fr.: February–May.
Sanskrit: Arani, Vijaya. Telugu: Thakkalaku, Thakkalomni.
Trade name: Wind killer, Glory tree.

Medicinal uses: The stem bark is given in dysentery and with other plants for unconsciousness after child birth. The leaves are used in fever due to sunstroke and malaria. The roots are used to cure Gonorrhoea.

Clerodendrum serratum (L.) Moon.

Family: Verbenaceae.

Erect shrubs, up to 2m tall, stem quadrangular. Leaves obovate or elliptic-oblong, glabrescent above, pubescent below. Flowers violet or lilac in terminal panicles. Fruit drupe, 4-lobed, black when ripe.

Occasional in forest areas of Telangana, Rayalaseema and north coastal districts.

Fl. & Fr.: August–December.

Propagation: by seeds.

Sanskrit: Bhriguja, Angaravallari. Telugu: Gantubarangi.

Medicinal uses: The root has a pungent, bitter and acrid taste, and is used in medicine for fevers, rheumatism and dyspepsia. The root decoction with ginger and coriander is taken for relieving nausea. The leaves are used for fevers and external applications in cephalagia and opthalmia. The seeds are slightly aperient and are used to some extent for dropsy.

Clitoria ternatea L.

Family: Leguminosae (Fabaceae)

Perennial twining shrubs, branchlets appressed tomentose. Leaves odd-pinnate, leaflets 2 or 3 pairs, opposite, ovate. Flowers bluish violet with yellow streaks or white, axillary, solitary. Fruit legume, flattened, beaked. Seeds 10–15, reniform.

Common on bushes along streams in forests, often cultivated.

Fl. & Fr.: September–February.

Sanskrit: Aparajitha. Telugu: Sankupuspam.

Trade name: Butterfly pea, Conchflower.

Medicinal uses: The leaves are diuretic, laxative, emetic and antiperiodic. The drug, shankupushi, mainly comprising the roots and seeds reported to be a nervine tonic, alterative and laxative. The leaves and roots are used in the treatment of a number of ailments including body aches, especially infections, urinogenital disorders and as anthelmintic and as antidotes to animal stings. Of the 2 varieties, the white flowered one is found to be therapeutically more active and hence preferred. The roots have an acrid and bitter taste, and are credited with purgative, laxative and diuretic properties and given in fevers. They are also demulcent and given in chronic bronchitis. The decoction or powder of root is given in rheumatism and ear-diseases. The seeds are laxatice and also cathartic but are considered safer for colic dropsy and enlargement of abdominal viscera, they are also used in swollen joints.

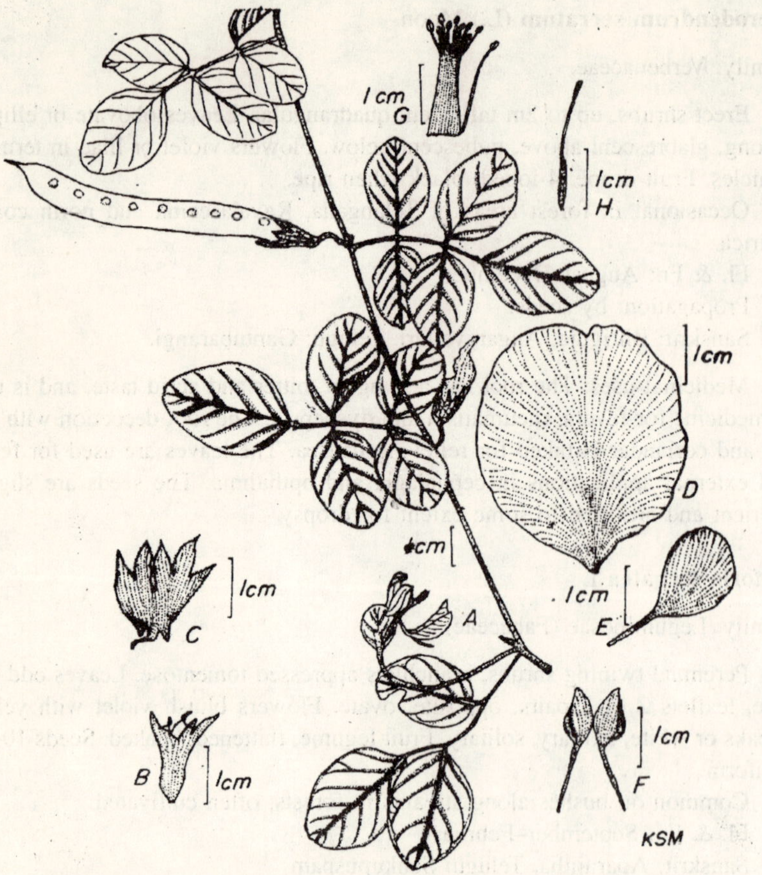

Fig. 42: *Clitora ternatea* L.
A. Twig; B. Calyx; C. Calyx split open; E. Wing petal; F. Keel petals; G. Staminal column; H. Pistil.

Coccinia indica Wight & Arn.

Syn: *Coccinia grandis* (L.) Voigt
Family: Cucurbitaceae.

Perennial herbaceous tendrillar climbers, tendrils slender, simple. Leaves entire or palmately lobed, glabrous. Flowers unisexual, white, solitary or subcymose, female flowers white, solitary. Fruit ovoid berry, red when ripe, many-seeded., seeds ovoid-oblong, compressed.

Common on hedges and bushes in all districts.
Fl. & Fr.: Through out the year.
Sanskrit: Bimbi, Bimbika. Telugu: Dondakaya.
Trade name: Ivy Gourd.

Medicinal uses: Various preparations of roots, stems and leaves of *Coccinia indica* have been mentioned in indigenous systems of medicine as being efficacious in the treatment of skin diseases, bronchial catarrh, bronchitis and diabetes.

The roots stems, leaves and fruits are used in indigenous system of medicine for diabetes.

Cochlospermum religiosum (L.) Alston

Syn: *Cochlospermum gossypium* DC.

Family: Cochlospermaceae.

Deciduous trees, up to 15 m tall, with yellow or red juice. Leaves 5-lobed, glabrous above, densely brownish tomentose below. Flowers large, golden yellow with silky sepals in terminal panicles. Fruit a large brown, leathery capsule, seeds numerous, woolly.

Common in dry forests especially on stony hills.

Fl.& Fr.: February–August.

Sanskrit: Girisalmalka, Silakarpasika. Telugu: Konda gogu, Konda Buruga.

Trade name: Yellow flowered cotton tree.

Propagation: seeds.

Medicinal uses: The gum from stem is sweetish, cooling, sedative and useful in coughs. Dried flowers and leaves are said to be used as stimulant.

Coffea arabica L.

Family: Rubiaceae.

Small ever green trees, up to 5 m tall. Leaves oblong-elliptic, opposite, glossy. Pointed at the tip. Flowers showy, fragrant, white, clustered in the leaf axils. Fruit round, mucilaginous, red when ripe. Seeds oval, grey green, enclosed in a silvery testa.

Cultivated in hills of Visakhapatnam and Vizianagaram districts.

Sanskrit: Rajpiluh. Telugu: Kaphi, Coffee.

Fl. & Fr.: March–October

Trade name: Coffee.

Propagation: by seeds.

Medicinal uses: Seeds stimulant, diuretic, used in whooping cough and hysteria. Coffeine is known as cerebral and respiratory stimulant. It is stimulant and facilitates digestion after heavy meal. It is a diuertic, reduces cutaneous perspiration. It is recommended in typhoid fever, intermittant fevers, periodic nervous asthma, whooping cough and albuminuria

Active principle: Alkaloid caffeine present in seeds.

Colocasia esculenta (L.) Schott.

Syn: *Colocasia antiquorum* Schott

Family: Araceae.

Robust tuberous herbs, tubers 1–5 cm across. Leaves large, 15–50 x 8–30 cm, ovate or suborbicular. Flowers in spadix. Female flowers many, arranged in close spirals, neuter flowers at the constriction of spathe, male flowers cream coloured. Fruit ovoid berry, many-seeded.

Common along canals, irrigation channels in most districts, also cultivated.

Fl. & Fr.: August–November.

Sanskrit: Alupam, Alukam. Telugu: Chemadumpa, Gedda.

Trade name: Taro, cocoyam.

Propagation: corms.

Medicinal uses: Pressed juice of the petioles is used as a styptic or astringent. The corms are rubifacient and laxative and used in otalgia, internal haemorrhages.

Convolvulus arvensis L.

Family: Convolvulaceae.

Annual, trailing or twining herbs. Leaves ovate-triangular, base hastate. Flowers rose-purple, solitary, axillary. Fruit globose capsule, glabrous, 4-seeded.

Occasional in forests in Nizamabad, Medak and Cuddapah districts.

Fl. & Fr.: July–February.

Sanskrit: Bhadravala, Rajabala.

Trade name: Deer's foot.

Propagation: by seeds.

Medicinal uses: The root possesses cathartic properties and purgative. The plant is frequently used to adultarate Sankhupushpi.

Corallocarpus epigaeus Clarke

Syn: *Bryonia epigaea* Rottl. & Willd.

Family: Cucurbitaceae.

Thick stemmed climbers, root stock large, tuberous, branches glabrous. Leaves broadly suborbicular, angled or deeply 3–5-lobed, densely villose. Flowers unisexual. Male flowers in crowded racemes on long axillary peduncles, female flowers usually solitary on short peduncles from the same axis. Fruit ellipsoid berry, glabrous, apex beaked, 6–9-seeded.

Common in most districts,

Fl. & Fr.: September–March.

Sanskrit: Pathalagaruda. Telugu: Nagadonda, Murudonda.

Medicinal uses: The root has a bitter and sub acid taste and is credited with alterative and laxative properties and is used in syphilitic rheumatism, later stages of dysentery and chronic mucous enteritis.

Cordia dichotoma Forst. f.

Syn: *Cordia obliqua* Willd.

Family: Boraginaceae.

Deciduous trees, up to 10m tall. Leaves broadly ovate. Flowers white to cream-coloured, in lax axillary and terminal cymes. Fruit ovoid drupe, cream coloured, mucilaginous, seed solitary.

Occasional in forests and hedges of fields.

Fl. & Fr.: December–January.

Sanskrit: Shleshmanthaka. Telugu: Bankiriki.

Medicinal uses: The fruit is mucilaginous, sweetish and edible. It is astringent, anthelmintic, diuretic, demulcent and expectorant. It is used in diseases of chest and urinary passage. A decoction of the bark is used in dyspepsia and fevers.

Coriandrum sativum L.

Family: Apiaceae (Umbellifereae).

Annual erect herbs, 30–45 cm tall. Leaves dimorphic, decompound, lower leaflets orbicular, deeply lobed, upper leaflets nearly sessile, bi or tripinnate with linear segments. Flowers white in terminal and leaf-opposed umbels, polygamous. Fruit globose or ellipsoid schizocarp.

Cultivated for seeds and foliage.

Fl. & Fr.: November–December.

Sanskrit: Dhanyaka. Telugu: Dhaniyalu, Kothimeera.

Trade name: Coriander.

Propagation: by seeds.

Medicinal uses: The fruits are considered carminative, diuretic, tonic, stomachic, antibilious, refrigerant and aphrodisiac. The leaves are used for the treatment of dyspepsia, flatulence and piles. Coriander leaves stimulate the appetite and act as a tonic. Corinader juice is recommended for patients who suffer from deficiency of Vitamin A,B and C. The paste of seeds ground with some water and applied to forehead can relieve headache.

Chemistry: Essential oil in seeds 0.3–2.2%. The essential oil is used for its flavor and carminative properties.

Costus speciosus (Koen.) Smith

Syn: *Banksea speciosa* Koen.

Family: Zingiberaceae-Costaceae.

Perennial herbs with creeping tuberous rhizomes. Leaves oblong-lanceolate. Flowers white with yellow tinge at throat in terminal globose spikes. Fruit globose capsule, seeds arillate.

Fl. & Fr.: June–November.

Sanskrit: Chanda, Pushkaramula. Telugu: Chengalva Kostu.

Trade name: Wild or spiral ginger

Propagation: by rhizomes, stem cuttings and seeds.

Medicinal uses: Rhizomes source of corticosteroids, sex hormones and used in birth control, tonic anthelmintic and purgative. The rhizomes are astringent, stimulant and anthelmintic. They are used to treat cough, catarrhal fever, dyspepsia, skin disease, worms and snake bite.

Chemistry: Rhizomes contain 2–4% Diosgenin. Diosgenin is the raw material for the synthesis of many steroidal drugs which include corticosteroids, sex-hormones, oral contraceptives and anabolic agents.

Crataeva magna (Lour.) DC.

Syn: *Crataeva nurvala* Buch-Ham.

Family: Capparaceae.

Moderate sized deciduous trees, up to 6m tall. Leaves digitately 3-foliolate, leaflets ovate or ovate-lanceolate. Flowers yellow in many-flowered lax terminal corymbs. Fruit ovoid berry.

Occasional in almost all districts.

Fl. & Fr.: April–July.

Sanskrit: Varuna, Ashmarighna, Ashmagna. Telugu: Vivapathri, Usikmanu.

Trade name: Three-leaved caper.

Propagation: by seeds.

Medicinal uses: The bark is bitter, antiperiodic, tonic and demulcent. An extract of bark is given as a laxative and for promoting appetite. It is also useful in calculus and other affections of the urinary organs. The root bark is rubifacient and counter irritant. The leaves are employed in poultices. The flowers are astringent and cholagogue.

Crossandra infundibuliformis (L.) Nees

Syn: *Justicia infundibuliformis* L.
Crossandra undulaefolia Salisb.

Family: Acanthaceae.

Under shrubs, up to 1.5 m tall. Leaves elliptic-oblong, glabrous. Flowers orange-scarlet in axillary and terminal dense spikes. Fruit capsule, 4-seeded.

Commonly cultivated in gardens for its ornamental flowers.

Fl. & Fr.: July–April.

Telugu: Kanakambaralu.

Trade name: Unarmed orange Nail dye.

Medicinal uses: Capsules are aphrodisiac.

Crotalaria laburnifolia L.

Family: Fabaceae (Papilionaceae)

Erect shrubs, up to 2.5m tall, branchlets slender, terete, glabrous. Leaves 3-foliolate, leaflets ovate. Flowers yellow in terminal or leaf-opposed racemes. Pod cylindric, 15–20-seeded.

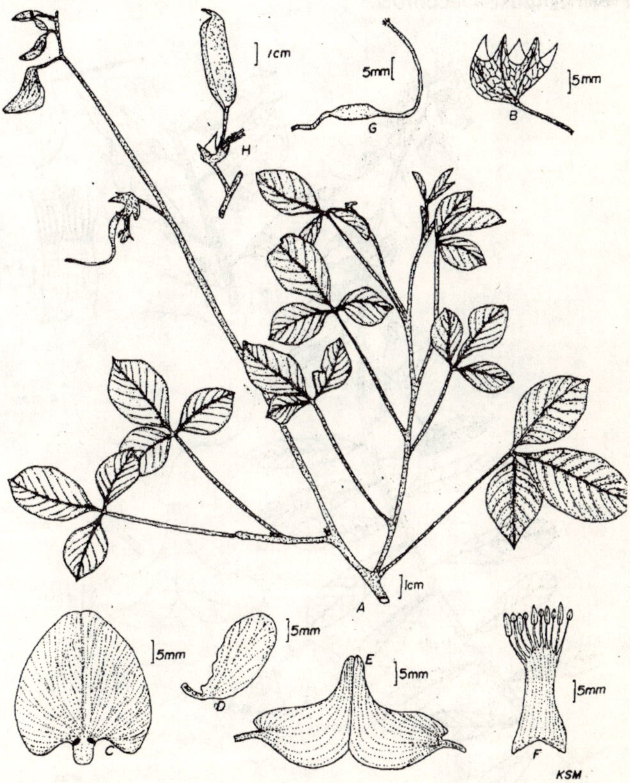

Fig. 43: *Crotalaria laburnifolia* L.
A. Twig; B. Calyx; C. Standard petal; D. Wing petal; E. Keel petals; F. Staminal column; G. Pistil; H. Pod.

Common in forests.

Fl. & Fr.: September–January.

Trade name: Pedda Giligicha.

Propagation: by seeds.

Medicinal uses: used for the scorpion and snake bites. Leaves infusion used as gargle.

Crotalaria retusa L.

Family: Fabaceae (Papilionaceae)

Undershrubs, up to 1.5 m tall, branchlets terete, pubescent. Leaves simple, oblanceolate. Flowers yellow in terminal racemes. Fruit pod, oblong, turgid 10–15-seeded.

Common weed in cultivated fields and waste lands.

Fl. & Fr.: August–December.

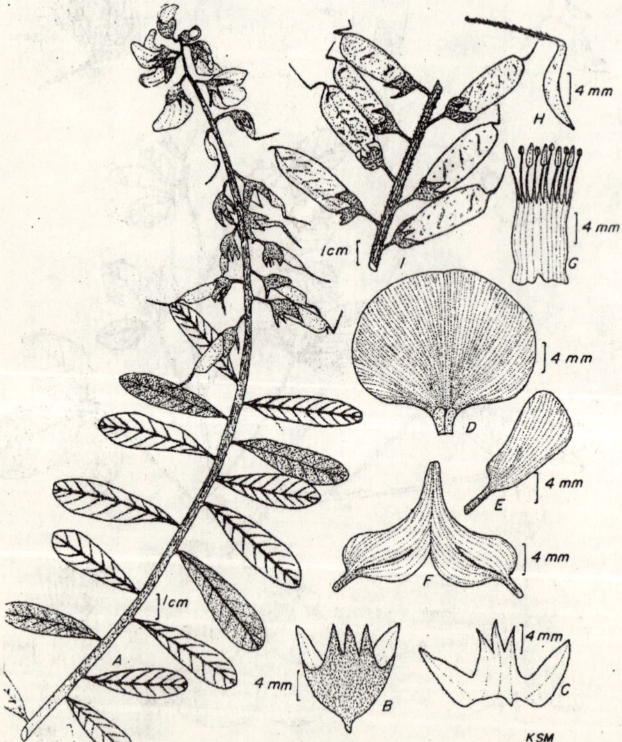

Fig. 44: *Crotalaria retusa* L.

A. Twig; B. Calyx; C. Calyx split open; D. Standard petal; E. Wing petal; F. Keel petals; G. Staminal column; H. Pistil; I. Fruiting twig.

Sanskrit: Sanapuspi, Sanaghantika. Telugu: Potti giligicha.
Trade name: Green yellow Sweet Pea.
Propagation: by seeds

Medicinal uses: The roots are used in chest diseases. The root crushed and mixed with spices is used as a remedy for colic. It is also used to treat haemoptysis. The leaves are used as a remedy for fever and a preperation of the plant is applied in skin diseases. Whole plant is astringent, expectorant, digestive and febrifuge. It is also used to treat Dyspepsia, fever, cardiac disorders, leprosy and flatulence.

Chemistry: The leaves contain indican. The seeds contain toxic alkaloid, monocrotaline.

Crotalaria verrucosa L.

Family: Fabaceae (Papilonaceae).

Erect subshrubs, up to 1m tall, branchlets 4-angled, appressed pubescent. Leaves simple, ovate or broadly rhomboid. Flowers bluish white in terminal racemes. Fruiț pod, oblong, cylindric, turgid.

Fig. 45: *Crotalaria verrucosa* L.
A. Twig; B. Calyx; C. Calyx split open; D. Standard petal; E. Wing petals; F. Keel petals; G. Staminal column; H. Pistil.

Common in open waste lands.
Fl. & Fr.: December–February.
Telugu: Giligicha.
Propagation: by seeds.

Medicinal uses: The leaves are used in the treatment of biliousness, dyspepsia, fever blood impurities, throat and mouth diseases and heart ailments. Juice of the leaves is also used in scabies and impetigo both internally and externally.

Cryptolepis buchanani Roem. & Schult.

Family: Asclepidaceae.

Large, much branched, climbing shrubs with plenty of latex. Leaves oblong to elliptic. Fruit paired follicles, basally inflated, pointed above, glabrous, seeds flat, silky hairy.
Common in forests.
Fl.& Fr.: March–April.
Sanskrit: Krishnasariba. Telugu: Adavipala teega.
Propagation: by seeds.

Medicinal uses: The root of the plant is bitter and aromatic. The root possesses demulcent, diaphoretic and diuretic properties. The preperation of the plant containing leaves and latex is given to children as a cure from rickets. It has cytotoxic glycosides and anticarcinogenic property. It is used for curing paralysis and rickets in children.

Chemistry: The roots and leaves contain cryptanosides C and A. Cryptanoside B and a Cardenolide Cryptosin have also been isolated from the leaves. Cryptosine is a cardioactive compound.

Cucumis sativus L.

Family: Cucurbitaceae.

Annual monoecious herbs, trailing or climbing, stems hirsute. Leaves broadly cordate-ovate, palmately 3–5-lobed, densely hairy. Flowers yellow, solitary, axillary. Fruits of various shapes, narrowly or broadly oblong, pale to dark-green or pale yellow, glabrous, many-seeded.
Cultivated for fruit vegetable.
Fl. & Fr.: Through out the year.
Sanskrit: Sukasa. Telugu: Dosakaya.
Trade name: Common cucumber.
Propagation: by seeds.

Medicinal uses: The seeds are reported to be cooling, tonic and diuretic. Fruits and seeds are used in Hyperdipsia, burning sensation, thermoplegia, fever, insomnia, bronchitis, general debility, constipation, renal calculus and urodynia.

Curculigo orchioides Gaertn.

Family: Hypoxidaceae.

Perennial acaulescent herbs, root stock stout, to 30 cm. Leaves lanceolate, prominently nerved, base sheathing. Flowers bright yellow in racemes or subcapitate clusters, lower ones bisexual, upper ones often male. Fruit oblong, succulent, 1–4-seeded.

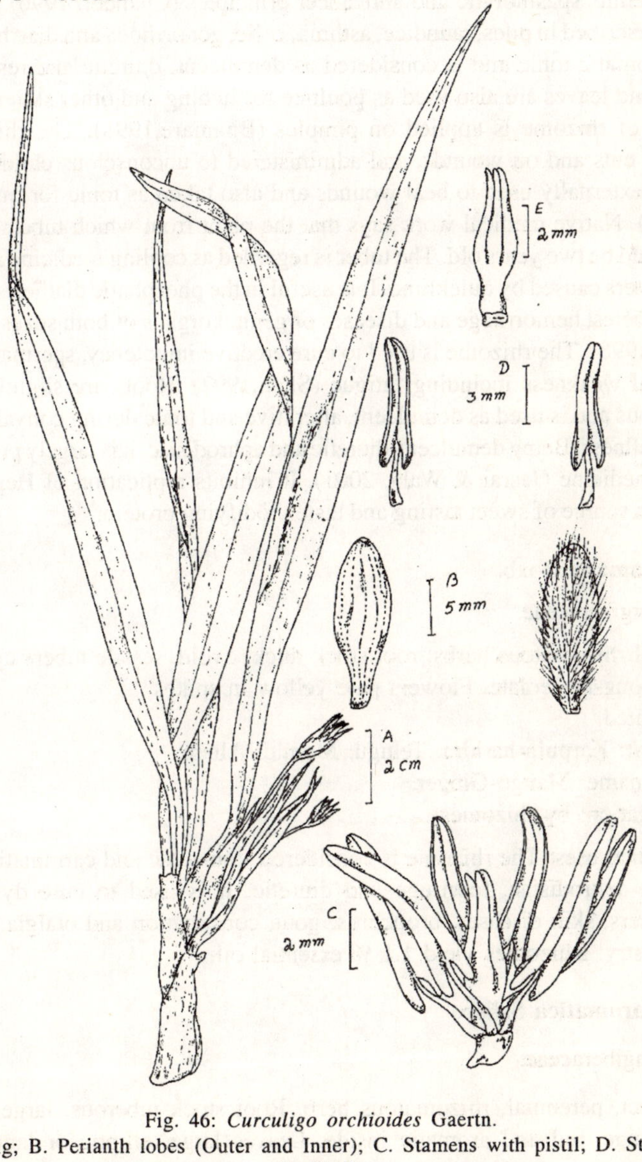

Fig. 46: *Curculigo orchioides* Gaertn.
A. Twig; B. Perianth lobes (Outer and Inner); C. Stamens with pistil; D. Stamens; E. Pistil.

Occasional in forest undergrowth in most districts.
Sanskrit: Talmulika. Telugu: Nelathatigedda.
Trade name: Black Musli.

Medicinal uses: Many medicinally important compounds are present in this species for which it is widely exploited and now endangered. The root, rhizome and leaves have medicinal properties. The whole plant is reported to have hypoglycaemic, spasmolytic and anticancer principles (Chandel, 1996). The rhizome is prescribed in piles, jaundice, asthma, colic, gonorrhoea and diarrhoea. It is a bitter aromatic tonic and is considered as demulcent, diuretic and restorative. Rhizome and leaves are also used as poultice for itching and other skin diseases. The paste of rhizome is applied on pimples (Bhamare,1998). The rhizome is applied on cuts and on wounds, and administered to unconscious children. The powder is externally used to heal wounds and also taken as tonic for impotency (Jain,1991). Native medical work says that the plant from which tubers are collected should be two years old. The tuber is regarded as cooling medicine and even to cure blisters caused by quicklime. It is useful in the phosphatic diathesis, scleroderma, diabetes, hemorrhage and diseases of genital organs of both sexes (Nair & Mohanan,1998). The rhizome is used to cure erective impotency, spermatorrhoea and general weakness including fatigue (Shah,1992). Roots are slightly bitter, mucilaginous and is used as demulcent, alterative and tonic during convalascence after acute illness. Being demulcent-diuretic and aphrodisiac, it is largely prescribed in native medicine (Jasrai & Wala, 2001). It inhibits replication of Hepatitis B virus. It is a source of sweet tasting and taste modifying protein.

Curcuma amada Roxb.

Family: Zingiberaceae.

Annual rhizomatous herbs, root stock thick, ovoid, sessile tubers cylindric. Leaves oblong-lanceolate. Flowers pale yellow in spikes.

Cultivated.
Sanskrit: Karpura haridra. Telugu: Mamidi Allam.
Trade name: Mango-Ginger.
Propagation: by rhizomes.

Medicinal uses: The rhizome is considered stomachic and carminative. It is used as an aphrodisiac, febrifuge and diuretic. It is used to cure dyspepsia, chronic ulcers, skin diseases, bronchitis, gout, constipation and otalgia.

Chemistry: Rhizomes yield 1.1 % essential oil.

Curcuma aromatica Salisb.

Family: Zingiberaceae.

An erect, perennial, rhizomatous herb. Root stock tuberous, large, tubers yellow, orange, red and aromatic inside. Leaves large, elliptic or lanceolate-oblong. Flowers pink in terminal racemose spike enclosed in coloured bracts. Fruit globose capsule.

Rare in East Godavari and Mahaboobnagar districts in forests.

Fl. & Fr.: June–September.

Sanskrit: Aranya Haridra, Vanaharidra. Telugu: Kasturi pasupu.

Trade name: Wild turmeric, Yellow Zedoary.

Propagation: by rhizomes.

Medicinal uses: The tubers are applied externally in combination with astringents, bitters and aromatics, to bruises and sprains. They are also used in skin eruption and infection and to improve the complexion.

Curcuma longa L.

Syn: *Curcuma domestica* Valeton

Family: Zingiberaceae

Perennial herb with thick underground rhizome. Leaves born in a tuft, broadly lanceolate. Flowers borne on a separate peduncle arising directly from the rhizome. Fruit capsule, seeds arillate.

Cultivated for its rhizome.

Sanskrit: Harida. Telugu: Pasupu.

Trade name: Turmeric.

Propagation: Rhizomes.

Medicinal uses: Curcumin possesses anti-microbial and anti-inflammatory properties.

In Indian systems of medicine tumeric is used to some extent as a stomachic, tonic and blood purifier, anthelmintic, antacid and carminative. It is also prescribed as an antiperiodic, alternative. Mixed with warm milk it is said to be beneficial in common cold. The juice of the fresh rhizome is used as an antiparasitic for many skin affections. A paste made from powdered shizome along with lime forms remedy for inflamed joints. It is also used to treat purulent ophthalmia.

Active principle: Curcumin-anti-inflamatory.

Curcuma pseudomontana Graham.

Syn: *Curcuma montana* sensu Baker

Family: Zingiberaceae.

Erect herbs, 75–80 cm tall with small root stock bearing subglobose tubers. Leaves green, 30–60 cm long, lanceolate-oblong. Flowers bright yellow in racemose spikes. Fruit 3-valved capsule, seed ovoid or oblong, usually arillate.

Rare in shady localities in Telangana and north coastal districts,

Fr. & Fr.: May–September.

Propagation: by rhizomes.

Telugu: Adavi Jongra.

Medicinal uses: It is reported to be used to treat leprosy, dysentery, cardiac diseases, haemorrhages, dysuria and general debility.

Cyamopsis tetragonaloba (L.) Taub.

Syn: *Cyamopsis psoralioides* DC.

Family: Fabaceae (Leguminosae).

Annual erect herbs, with appressed hairs. Leaves tripinnate, leaflets broad-elliptic, pubescent. Flowers small, purplish in axillary racemes. Fruit legume, 3–5cm long.

Cultivated for its fruit vegetable and seeds.

Telugu: Matikkayalu, Chowlekayalu, Goruchikkudu.

Trade name: Guar Gum, cluster bean.

Medicinal uses: Gum is used as a binder and disintegrater in tablets. Gum is also employed as suspensions or thickener in emulsions. Guar gum is also used for curing night blindness, dyspeptic complaints, anorexia, constipation and agalactia.

Active principle: Galactomannans, saponins.

Cyclea peltata (Lam.) Hook.f.& Thoms.

Syn: *Menispermum peltatum* Lam.
Cyclea burmanni Hook.f.& Thoms.

Family: Menispermaceae.

Twining pubescent herb. Leaves cordate, coriaceous. Flowers unisexual, in axillary panicles, male flowers regular, female flowers irregular. Fruit obovoid drupe.

Rare in Visakhapatnam and Chittor districts.

Fl. & Fr.: January–April.

Sanskrit: Durva, Patha.

Medicinal uses: It is known in trade as pathamool and pathaa. Roots are reported to be useful in treating cough, bronchitis, colic, diorrhoea, dysentery, dropsy, painful swellings leprosy, skin diseases, ulcers, wounds vomiting and cardiac disorders. The leaves are cooling and reported to be useful in dandruff, burning sensation of the eye and fever.

Cymbopogan flexuosus (Nees ex Steud.) Watson

Syn: *Andropogon flexuosus* Nees ex Steud.

Family: Poaceae (Gramineae).

A perennial tufted grass, up to 3 m tall. Leaf sheaths 6–15 cm long, glabrous, blade lanceolate. Spikelets in lax panicles.

Common among hill tops and open forests.

Fl. & Fr.: July–December

Telugu: Nimma gaddi.

Trade name: Lemon grass.

Propagation: by slips.

Medicinal Uses: Leaves are source of Vitamin A. Leaves are used to cure leprosy and epilepsy. The leaves are also used as mosquito repellant and in creams.

Cymbopogon martinii (Roxb.) Watson

Syn: *Andropogon martinii* Roxb.

Family: Poaceae (Gramineae)

A perennial tufted, sweet scented grass up to 2 m tall. Leafsheaths 7–15 cm long, Blades linear-lanceolate. Spikelets in panicles.

Occasional in forests.

Fl. & Fr.: July–December.

Sanskrit: Rohisah, Dhyama Kah. Telugu: Kamakshi Kasuvu.

Trade name: Palma Rosa, Rosha grass.

Propagation: by slips.

Medicinal uses: Leaves are cardiotonic, galactogogue, sudorific, febrifuge, neuralgia, bronchitis, leprosy and epileptic fits.

Chemistry: Leaves contain 1.4% oil (dry basis)

Cymbopogon winterianus. Jowitt

Family: Poaceae (Gramineae)

Tuted, aromatic, perennial herbs, culms erect, leaf blades linear. Flowers in terminal decompound panicle.

Cultivated.

Trade name: Java citronella, Burma Citronella.

Medicinal uses: Leaves used as deodorant and mosquito repellant.

Cyperus rotundus L.

Family: Cyperaceae.

Perennial, glabrous, erect, stoloniferous herb, stolons bearing ovoid, tunicate, black fragrant tubers. Leaves all radical, narrowly linear, flat, 10–30 cm long, sheath brown, spikelets brownish in compound umbels, bracts 3–5. Fruit oblong, trigonous achene, greyish black.

Common weed of cultivated fields.

Fl. & Fr.: October–December.

Sanskrit: Musta, Hima, Kakshottha. Telugu: Tunga.

Trade name: Nut grass, Cocograss.

Medicinal uses: Tubers are reported to be used in conjunctivitis (oil, topical), diarrhoea, indigestion, deodorant, cholera, dysentery, fever, inflammation, leprosy, skin diseases, scabies, amnesia, epilepsy, anorexia, dyspepsia, flatulence, colic, verminosis and general debility.

Active principle: 0.5% essential oil and unstable alkaloid.

Dalbergia latifolia Roxb.

Family: Fabaceae (Leguminosae).

Medium sized deciduous trees, up to 10 m tall, branches terete, glabrous. Leaves odd-pinnate, leaflets 7–9, alternate, obovate-orbicular. Flowers white with pale yellow tinge on old wood. Fruit legume, oblong-lanceolate, 1–3-seeded.

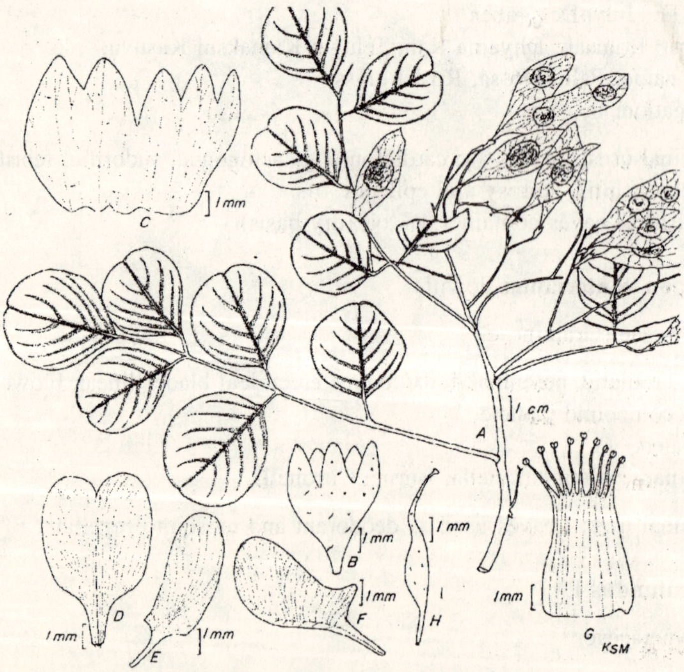

Fig. 47: *Dalbergia latifolia* Roxb.

A. Twig; B. Calyx; C. Calyx split open; D. Standard petal; E. Wing petal; F. Keel petal; G. Staminal coumn; H. Pistil.

Common in forests in all districts.
Fl. & Fr.: July–February.
Sanskrit: Sispa. Telugu: Jittegi; Irugudu.
Trade name: Rose wood.
Propagation: seeds.

Medicinal uses: Parts of the tree are reported to be useful as stimulant and appetiser and also in dyspepsia, diarrhoea, leprosy, obesity and worms. Bark is also reported to be used for bodyache, ulcers, wounds, eczema and pimples.

Dalbergia sissoo Roxb.

Family: Fabaceae (Leguminosae).

A medium sized deciduous tree, up to 30m tall. Leaves imparipinnate, leaflets 3–5, alternate, obovate-suborbicular, glabrous. Flowers creamish in terminal panicles.

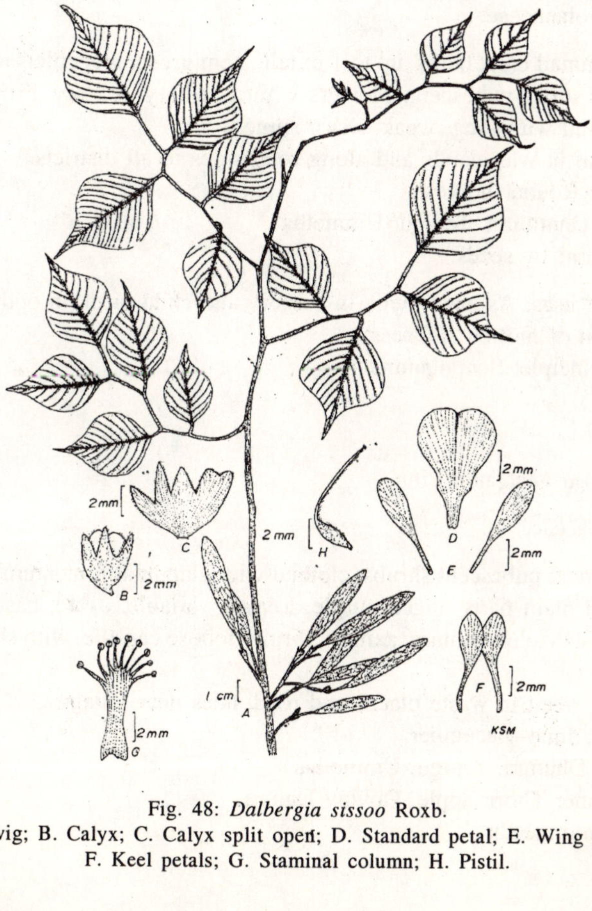

Fig. 48: *Dalbergia sissoo* Roxb.
A. Twig; B. Calyx; C. Calyx split open; D. Standard petal; E. Wing petals; F. Keel petals; G. Staminal column; H. Pistil.

Fruit oblong legume, glabrous, 1–2-seeded.
Planted as avenue tree.
Fl.& Fr.: January–September.
Sanskrit: Simsapa. Telugu: Sissoo.
Trade name: Sissoo.
Propagation: by seeds.

Medicinal uses: The wood is used in the treatment of scabies and leprosy. The leaves are used for the cure of eye diseases. The leaves of sissoo are bitter and stimulant. A decoction of the leaves is said to be useful in gonorrhoea. The leaf mucilage mixed with sweet oil is applied in excoriations. The roots are astringent and are useful in diarrhoea and dysentery.

Datura innoxia Mill.

Syn: *Datura metel* auct.non Linn.

Family: Solanaceae.

Stout stemmed erect herbs, up to 1 m tall, stem green, branchlets tomentose. Leaves ovate, softly pubescent. Flowers white, axillary, solitary. Fruit globose capsule, covered with long, weak, sharp spines.
Occasional in wastelands and along road sides in all districts.
Fl. & Fr.: Round the year.
Sanskrit: Dhatturah. Telugu: Ummetha
Propagation: by seeds.

Medicinal uses: As anaesthetic in surgery and child birth, in opthalmology and prevention of motion sickness.
Active principle: Seopolamine.

Datura metel L.

Syn: *D. fastuosa* auct. non Linn.

Family: Solanaceae.

Annual erect, pubescent shrubs, clothed with glandular tomentum, up to 1m tall. Stem and plant parts tinged purple. Leaves variable, ovate, base unequal. Flowers pinkish-violet, solitary, axillary. Fruit globose capsule, with short, stout, blunt spines.
Common weed in waste places and road sides near villages.
Fl. & Fr.: June–December.
Sanskrit: Dhatura. Telugu: Ummetha.
Trade name: Thorn apple, Downy Datura.
Propagation: seeds.

Medicinal uses: Contains scopolamamine used as anticholinergic.

The dried leaves, flowering tops and seeds constitute the drug. The chief active principle in the leaves is hyoscyamine. The drug is therefore, useful in the same manner as Belladonna or Hyoscyamus. The drug is useful in Bronchitis or asthma. It is antispasmodic and narcotic.

Decalepis hamiltonii Wight & Arn.

Family: Asclepiadaceae.

Large glabrous climbing shrub, branches jointed, branchlets terete. Leaves obovate-orbicular, glabrous. Flowers white, small in axillary peduncled, trichotomous cymes. Fruit follicle, oblong-lanceolate.

Rare in forests of Kurnool, Chittor, Nellore, Anantapur and Cuddapah districts.

Fl.: April–September. Fr.: September–February.

Sanskrit: Sariba, Sweta Sariva.

Telugu: Nannari, Maredu kommulu.

Propagation: by seeds.

Medicinal uses: Roots are used as flavouring agent, appetizer and blood purifier. Root extract was found to be a potent antimicrobial agent. The highly aromatic roots are used as a culinary spice and in preparing popular cool drink locally known as 'Nannari' which has cooling effect in summer.

Delonix elata (L.) Gamble

Syn: *Poinciana elata* L.

Family: Leguminosae-Caesalpiniaceae.

Trees, up to 12 m tall, branchlets warty puberulous. Leaves bipinnate, pinnae 6–9 pairs, leaflets 10–22 pairs, oblong-elliptic. Flowers yellowish-white in terminal racemes. Fruit pod, dark brown.

Planted as an avenue tree.

Fl. & Fr.: December–June.

Trade name: White Gold Mohur.

Propagation: by seeds, stem cuttings.

Medicinal uses: Leaves are said to be used in swellings, inflammations, rheumatism, flatulence and as laxative. Bark is considered febrifuge and antiperiodic.

Delonix regia (Boj. ex Hook.) Raf.

Syn: *Poinciana regia* Boj. ex Hook.

Family: Caesalpiniaceae-Leguminosae

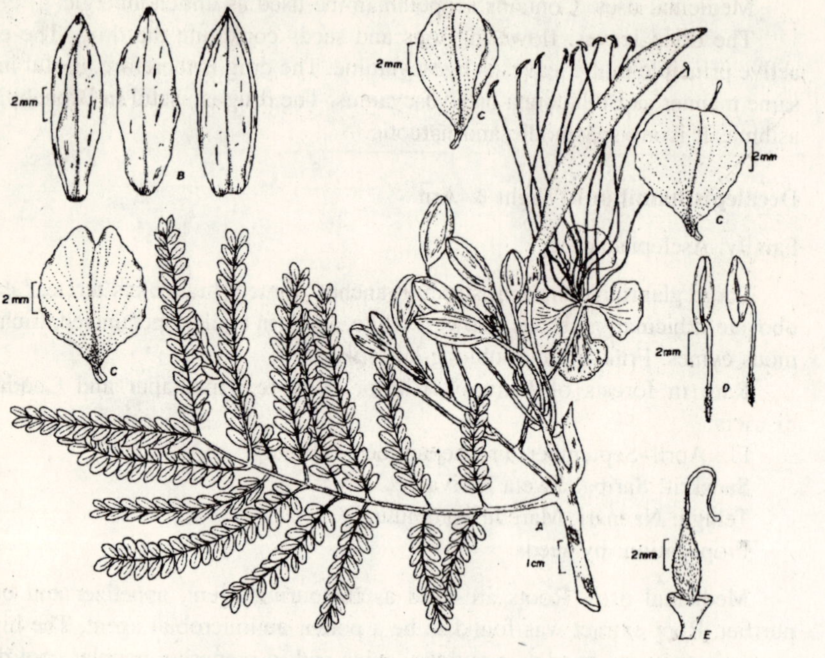

Fig. 49: *Delonix elata* (L.) Gamble
A. Twig; B. Sepals; C. Petals; D. Stamens; E. Pistil.

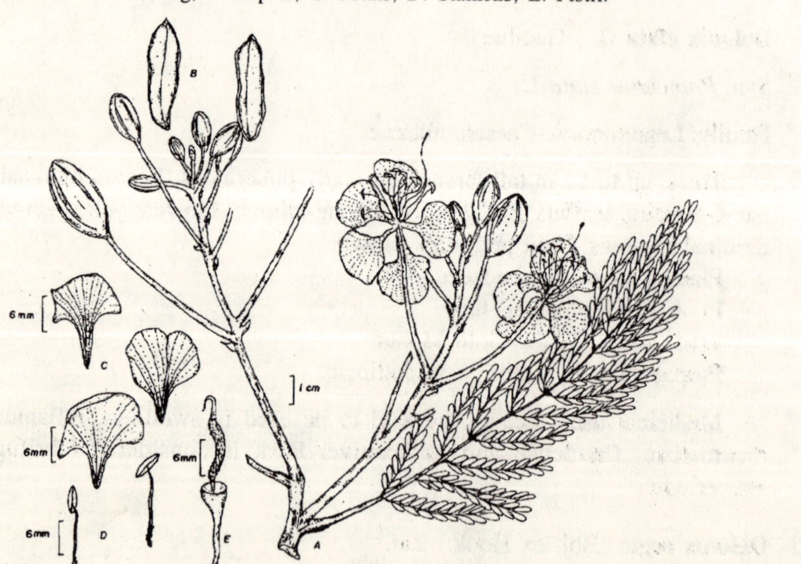

Fig. 50: *Delonix regia* (Boj.ex. Hook.) Rafin.
A. Twig; B. Sepals; C. Petals; D. Stamens; E. Pistil.

Deciduous trees, with grey to pale-brown bark, up to 12.m tall. Leaves 2-pinnate, pinnae 8–20 pairs, leaflets 12–30 pairs, oblong. Flowers red in simple or branched racemes. Pod broadly linear, woody, dark brown or reddish brown, flat.

Planted in gardens and along road sides.
Fl. & Fr.: April–August.
Telugu: Sima Sunkesula, Agnipoolu, Erraturayi.
Trade name: Flamboyant flame tree, Gulmohar.
Propagation: by seeds, by stem cuttings.

Medicinal uses: Flowers are used in dysmenorrhoea.

Desmodium gangeticum (L.) DC.

Syn: *Hedysarum gangeticum* L.

Family: Fabaceae (Leguminosae).

Erect under shrubs, up to 1m tall, branches angled, grooved, sparsely pubescent. Leaves simple, membranous, oblong, obovate or lanceolate. Flowers deep violet or white in axillary and terminal panicles. Fruit linear pod, moniliform, 6–8-jointed.

Occasional in forests in most districts.
Fl. & Fr.: August–January.
Sanskrit: Shalaparni. Telugu: Gitanaram.
Propagation: seeds.

Medicinal uses: The root of the plant, known in trade as 'Shaalparnimool' is common ingredient in Ayurvedic preperations. It is used in diarrhoea, chronic fever, biliousness, cough and vomitting. It is also used in treating pyorrhoea. The seeds are powdered and the paste is used in skin eruptions.

Desmodium motorium (Houtt.) Merr.

Syn: *Hedysarum motorium* Houtt.
Desmodium gyrans (L.f.) DC.

Family: Fabaceae (Papilionaceae.)

Erect undershrubs, up to 1 m tall, branchlets slender. Leaves 3-foliolate, leaflets linear-lanceolate. Flowers pink in terminal or axillary racemes. Fruit pod, oblong, joints up to 9.

Occasional in forests.
Fl. & Fr.: September–February.
Trade name: Indian Telegraphic plant.
Propagation: by seeds.

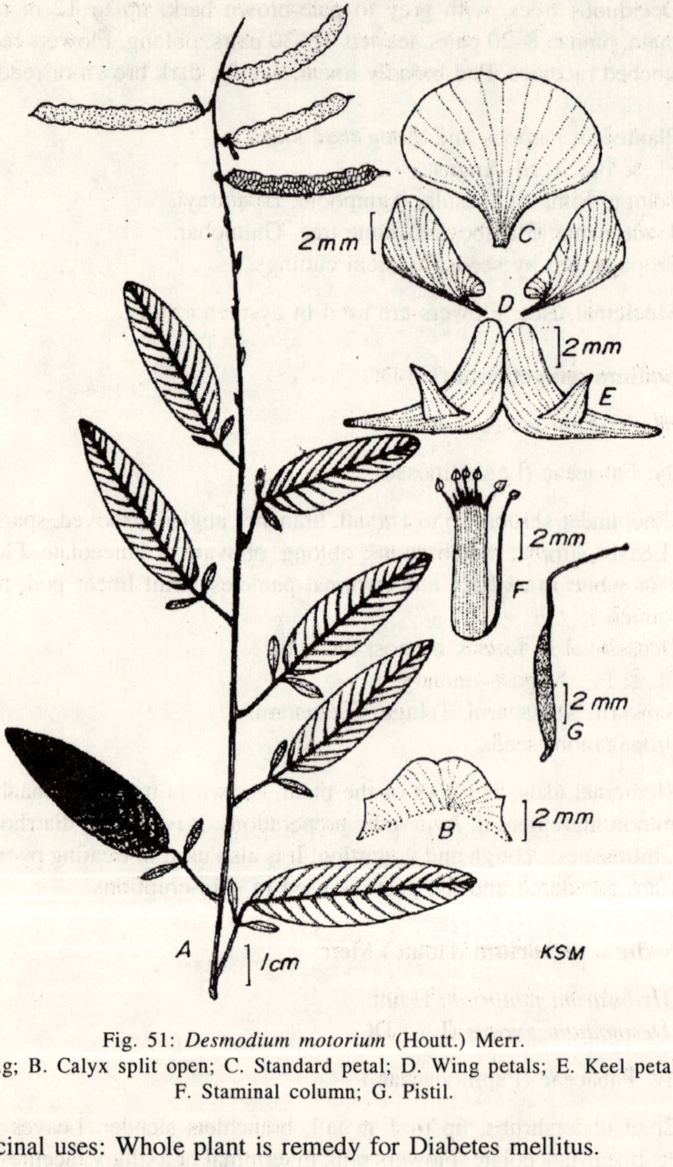

Fig. 51: *Desmodium motorium* (Houtt.) Merr.
A. Twig; B. Calyx split open; C. Standard petal; D. Wing petals; E. Keel petals; F. Staminal column; G. Pistil.

Medicinal uses: Whole plant is remedy for Diabetes mellitus.

Desmodium triflorum (L.) DC.

Syn: *Hedysarum triflorum* L.

Family: Fabaceae (Papilionaceae)

Prostrate, spreading herbs, branches rooting at nodes. Leaves trifoliolate,

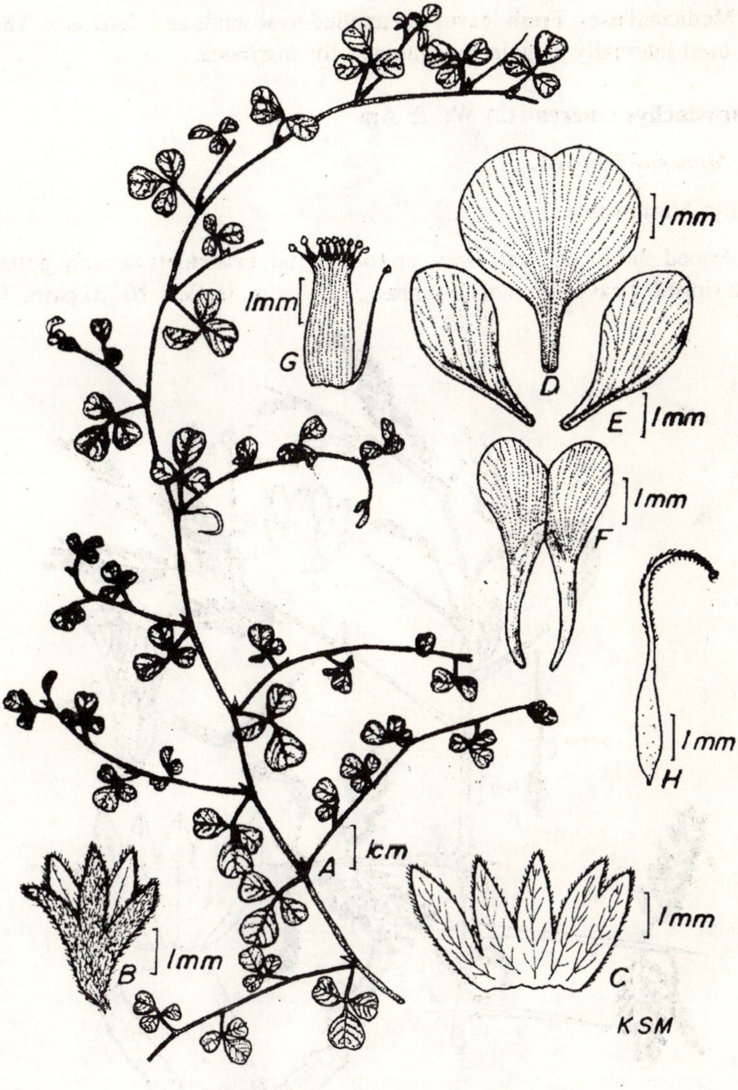

Fig. 52: *Desmodium triflorum* (L.)DC
A. Twig; B. Calyx; C. Calyx split open; D. Standard petal; F. Keel petals; G. Staminal column; H. Pistil.

leaflets obovate. Flowers purplish pink or pinkish blue, 2–5 clustered in axillary fascicles. Fruit pod, jointed.

Common in hedges and along rivers.

Sanskrit: Tripadi. Telugu: Munta Mandu.

Propagation: by seeds.

Medicinal uses: Fresh leaves are applied to wounds and abscesses. They are also used internally as galactogogue and for diarrhoea.

Dichrostachys cinerea (L.) Wt. & Arn.

Syn: *Mimosa cinerea* L.

Family: Mimosaceae.

Armed shrubs or small trees, up to 6 m tall, branchlets densely pubescent, spine tipped. Leaves bipinnate, pinnae 5–10 pairs, leaflets 10–20 pairs, linear-

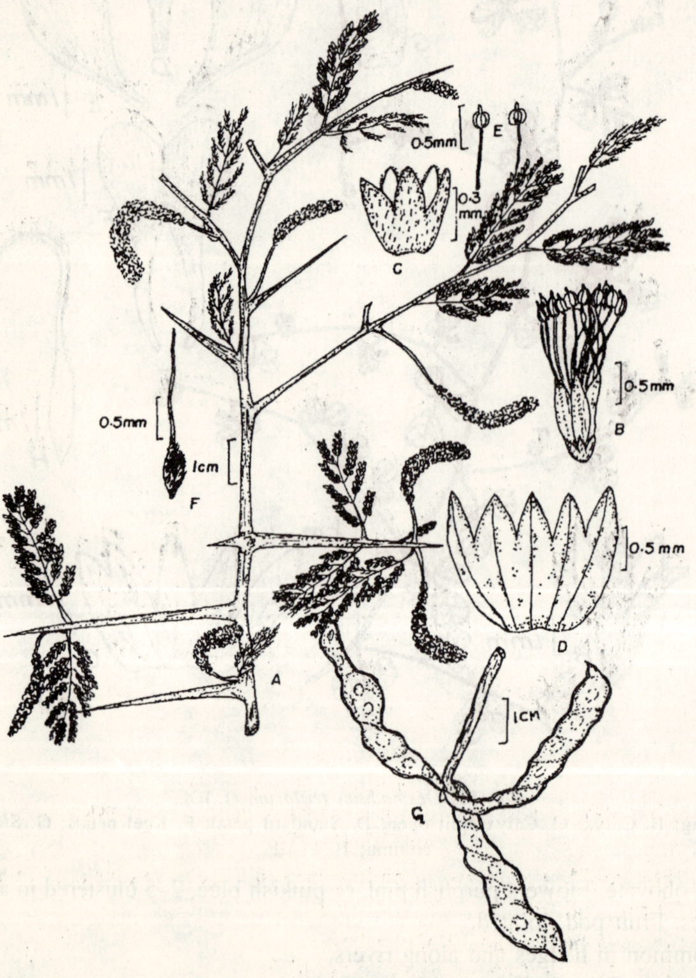

Fig. 53: *Dichrostachys cinerea* (L.) Wight & Arn.
A. Twig; B. Flower; C. Calyx entire; D. Corolla split open; E. Stamens; F. Pistil; G. Pod.

oblong. Flowers in cylindric peduncled spikes, spikes dimorphic, bicoloured, upper yellow flowers bisexual, lower sterile with large staminodes. Fruit linear pod, coiled when ripe, 4–6-seeded.

Common in scrub jungles and open forests.
Fl. & Fr.: May–December.
Sanskrit: Vellantara. Telugu: Velthuri.
Propagation: by seeds.

Medicinal uses: Tender shoots of the plant are bruised and applied to the eyes in cases of opthalmia. The root is astringent and used in rheumatism, urinary calculi and renal troubles. The roots are also used in inflammation, arthralgia, elephantiasis, dyspepsia, diarrhoea, strangury, vaginopathy, metropathy, paralysis and stones in the kidney.

Dioscorea alata L.

Syn: *Dioscorea atropurpurea* Roxb.
Dioscorea globosa Roxb.
Dioscorea purpurea Roxb.

Family: Dioscoreaceae.

A large climber, up to 15 m long, with stems twining to the right. Leaves opposite or rarely alternate, usually 5-nerved. Bulbils globose, ovoid. Tubers brown to black in colour, non poisonous, edible.

Cultivated in the north coastal districts.
Fl. & Fr.: September–December.
Sanskrit; Pindalu. Telugu; Pendalamu.
Trade name: Greater yam, Asiatic yam.

Medicinal uses: The tuber is considered anthelmintic and useful in leprosy, piles and gonorrhoea.

Dioscorea bulbifera L.

Syn: *Dioscorea sativa* Thunb.non Linn.

Family: Dioscoreaceae.

A vine. Stems twining to the left, bulbiferous. Roots tuberous, globose. Leaves broadly ovate-sub-orbicular. Flowers unisexual, male flowers green or purplish, female flowers in axillary spikes. Fruit capsule, recurved.

Common in dry deciduous forest.
Fl. & Fr.: September–December.
Telugu: Adavi dumpa.
Trade name: Potato yam.
Contains Diosgenin 4% in the tuber.

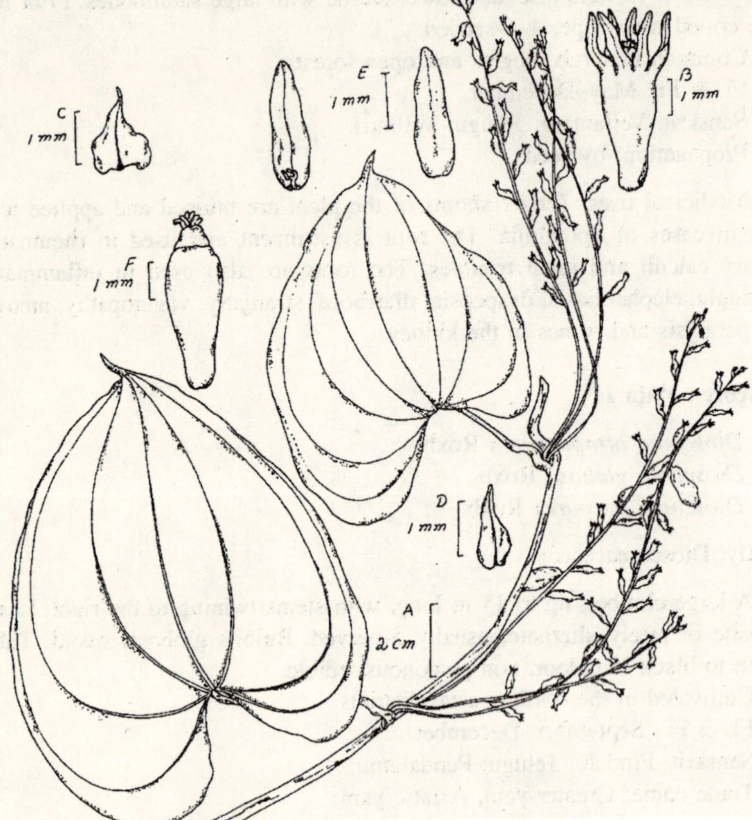

Fig. 54: *Dioscorea bulbifera* L.-Female.
A. Twig; B. Female flower; C. Bract; D. Bracteole;
E. Perianth lobe (Dorsal and Ventral); F. Pistil.

It has diuretic and anti-inflammatory activity (Dhawan et al., 1977), and is reported to be one of the major Indian medicinal plants used in three indigenous systems of medicine (Kamboj, 2000).

Dodonaea viscosa (L.) Jacq.

Family: Sapindaceae

Gregarious evergreen shrubs. Leaves simple, elliptic-oblanceolate. Flowers greenish-yellow, polygamous in lateral or terminal branched racemes. Fruit 2–3-winged capsule, seeds black.

Common in scrub jungles in all districts.

Fl. & Fr.: August–January.

Sanskrit: Aliar, sanatta. Telugu: Bandaru.
Trade name: Jamaica Switch Sorrel.

Medicinal uses: Leaves are used for setting broken bones.

Drimia indica (Roxb.) Jessop.

Syn: *Urginea indica* (Roxb) Kunth.

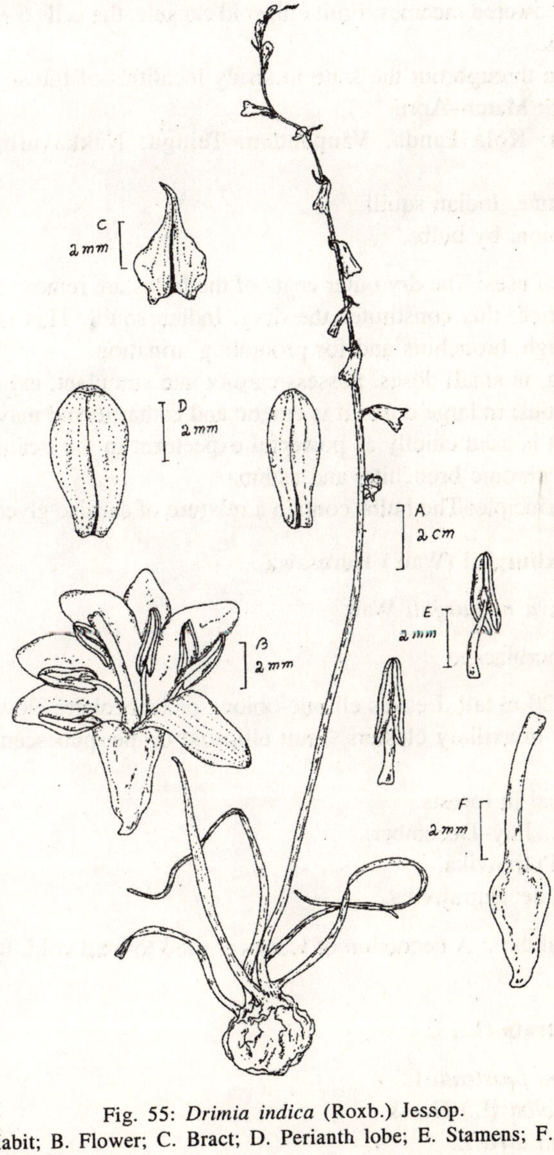

Fig. 55: *Drimia indica* (Roxb.) Jessop.
A. Habit; B. Flower; C. Bract; D. Perianth lobe; E. Stamens; F. Pistil.

Urginea coromandeliana (Roxb.) Hook.f.
Urginea wightiana Hook.
Scilla coromandeliana Roxb.

Family: Liliaceae.

A herb with tunicate bulb, bulb the size of an apple, pale, ovoid, thick. Leaves appearing after the flowers, nearly flat, linear. Flowers dingy brown in slender, lax flowered racemes. Fruit ellipsoid capsule, the cells 6–9-seeded, seeds elliptic, black.

Common through out the state in shady localities of forest.

Fl. & Fr.: March–April.

Sanskrit: Kola kanda, Vanpandan. Telugu: Nakkavulligadda, Adavi Tellagadda.

Trade name: Indian squill.

Propagation: by bulbs.

Medicinal uses: The dry outer coats of the bulbs are removed, the bulbs are sliced and dried; this constitutes the drug, Indian squill. This is used in heart ailments, cough, bronchitis and for promoting urination.

The drug, in small doses, possess cardiotonic stimulant, expetorant and diuretic properties; in large doses it is emetic and cathartic and may cause cardiac depression. It is used chiefly as powerful expectornt in the treatment of cough, especially in chronic bronchitis and asthma.

Active principle: The bulbs contain a mixture of cardiac glycosides (O.3%).

Drypetes roxburghii (Wall.) Hurusawa

Syn: *Putranjiva roxburghii* Wall.

Family: Euphorbiaceae.

Tree, to 20 m tall. Leaves elliptic-oblong or lanceolate. Flowers unisexual, male flowers in axillary clusters. Fruit ellipsoid drupe, pubescent, seed crustaceous.

Occasional in forests.

Fl. & Fr.: July–December.

Telugu: Putrajivika.

Trade name: Putrajivika.

Medicinal uses: A decoction of leaves is used to treat cold, fever and rheumatism.

Eclipta prostrata (L.) L.

Syn: *Verbesina prostrata* L.
Eclipta alba (L.) Hassk.
Verbesina alba L.

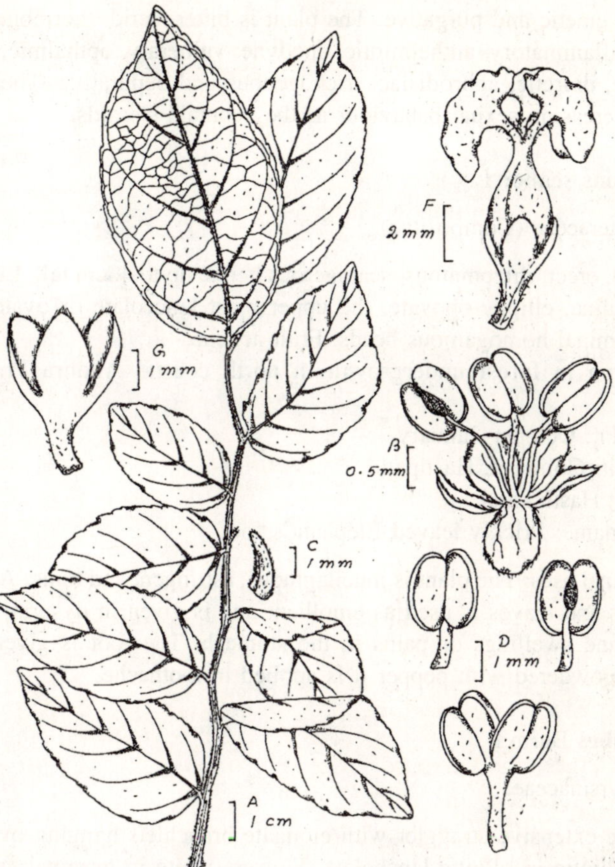

Fig. 56: *Drypetes roxburghii* (Wall.) Hurusawa
A. Female flower; B. Bract; C. Stamens; D. Anthers male; E. Twig; F. & G. Tepals.

Family: Asteraceae.

Annual prostrate herb with ascending branches, to 70 cm tall. Leaves oblong-lanceolate. Flowers in terminal or axillary heterogamous heads, white. Fruit achene, obovate-oblong, triquetrous.

Common in moist localities, paddy fields and near streams and ponds.

Sanskrit: Bhringaraj. Telugu: Guntagalagara.

Trade name: Trailing Eclipta.

Medicinal uses: The herb is used as tonic and deobstruent in hepatic and spleen enlargements and in skin diseases. The plant juice is administered in combination with aromatics for catarrhal jaundice. It is also reported to be effective in blackening and strengthening of the hair and in strengthening of gums.

The root is emetic and purgative. The plant is bitter, acrid, thermogenic, alterative, antiinflammatory, anthelmintic, anodyne, vulnerary, opthalmic, digestive, carminative, diuretic, aphrodisiac, trichogenous and depurative. The seeds are good for increasing sexual behaviour as they contain steroids.

Elephantopus scaber L.

Family: Asteraceae (Compositae).

Annual, erect, rhizomatous, scapigerous herbs, up to 40cm tall. Leaves radical and cauline, elliptic-obovate, the upper most lanceolate or ovate. Flowers pink, in terminal homogamous heads. Fruit achene.

Occasion in forest undergrowth in north coastal Andhra Pradesh and Telangana.

Fl. & Fr.: October–January.

Sanskrit: Gojihava, Hastipadi.

Telugu: Hasthikaraka.

Trade name: Prickly leaved Elephant's foot.

Medicinal uses: The plant is mucilaginous, astringent and tonic. A decoction of the roots and leaves is used as emollient and is given in dysuria, diarrhoea, dysentery and swellings or pains in the stomach. The root is given to arrest vomitting, powdered with pepper it is applied in toothache.

Embelia ribes Burm.f.

Family: Myrsinaceae.

A large extensive straggler with elongate branchlets hanging over. Leaves alternate, elliptic or elliptic-lanceolate. Flowers white in terminal and axillary panicle. Fruit globose drupe, 1-seeded.

Rare in Visakhapatnam district.

Fl. & Fr.: March–August.

Sanskrit: Vidang. Telugu: Vayividangamu.

Trade name: Embelia.

Medicinal uses: Root infusion is given for coughs and diarrhoea. decoction of dried fruits is used for fevers and diseases of chest and skin. Dried fruit is anthelmintic, astringent, tonic; used in scorpion sting and snake bite. Leaves are useful in pruritus, skin diseases and leprosy.

Embelia tsjeriaṃ-cottam DC.

Syn: *Embelia rọbusta* auct.

Family: Myrsinaceae.

A rambling shrub or small tree; young branchlets rusty tomentose. Leaves alternate, broadly elliptic, gland-dotted. Flowers whitish in colour, in axillary rusty-puberulous racemes. Berry globose, 1-seeded.

Rare in forests of coastal and Rayalaseema districts.

Fl. & Fr.: April–July.

Sanskrit: Bidanga, Vidanga.

Propagation: by seeds.

Medicinal uses: The uses of *Embelia tsjeriam-cottam* are similar to *E. ribes* and they are used as a substitute. Fruits are laxative and effective in the treatment of tape worms.

Emilia sonchifolia (L.) DC.

Syn: *Cacalia sonchifolia* L.

Family: Asteraceae (Compositae).

Annual, erect herbs, to 40 cm tall. Leaves radical and cauline, lower ones lyrately lobed, spathulate. Flowers pink in terminal homogamous heads. Fruit achene, angled, 5-ribbed.

Occasional in moist places.

Fl. & Fr.: August–February.

Sanskrit: Sudhirmudi.

Propagation: by seeds.

Medicinal uses: The plant is sudorific. A decoction of it is used as febrifuge in infantile tyanpanites and in bowel complaints. The root is used for diarrhoea. the juice of fresh leaves is used for sore ears, sore eyes and night-blindness. The plant is used also for cuts and wounds.

Enicostemma axillare (Lam.) Roynal

Syn: *Enicostemma hyssopifolium* (Willd.) Verdc.
Enicostemma littorale auct.
Exacum hyssopifolium Willd.

Family: Gentianaceae.

Perennial, erect, glabrous herb, up to 40m tall, stem 4-angular. Leaves decussate, lanceolate-oblong. Flowers white, sessile, in dense, axillary, clusters. Fruit septicidal capsule, 2-valved.

Occasional in moist places.

Fl. & Fr.: June–January.

Sanskrit: Nagaguha. Telugu: Nelagolimidi, Nelagurugudi.

Trade name: Indian Gentian.

Propagation: by seeds.

Medicinal uses: The plant is bitter, tonic, stomachic and laxative. The whole plant is dried, powdered and given with honey as a blood purifier and in dropsy, rheumatism, abdominal ulcers, hernia, swellings, itches and insect poisoning. It is used as a substitute for chiretta (*Swertia chirayita*). The plant is bitter due to presence of glycosides and ophelic acid. It is acrid, antiinflammatory, antipyretic and a liver tonic.

Entada pursaetha DC.

Syn: *Entada scandens* auct non Benth.

Mimosa entada L.

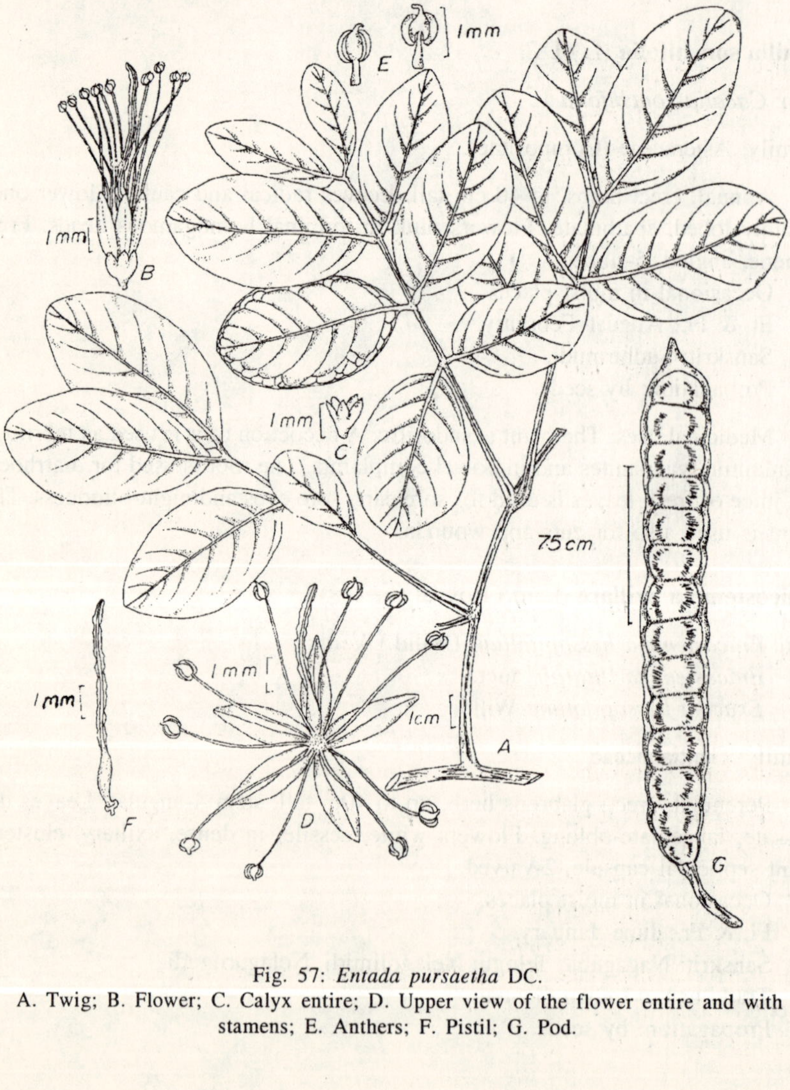

Fig. 57: *Entada pursaetha* DC.

A. Twig; B. Flower; C. Calyx entire; D. Upper view of the flower entire and with stamens; E. Anthers; F. Pistil; G. Pod.

Family: Mimosaceae.

Unarmed lianas, up to 30 m long. Leaves alternate, bipinnate, pinnae 2 pairs, leaflets about 4 pairs, ovate-obovate, main rachis generally ending in a bifid tendril. Flowers pale green panicled spikes. Pods elongate, flat, up to 1 m long, woody, falcate, indented between seeds, seeds orbicular, shinig.

Hills forests Nallamalais, Chittoor and coastal districts.

Fl. & Fr.: March–July.

Telugu: Gilatiga.

Trade name: Elephant creeper, Mackay bean, Garbee bean.

Medicinal uses: The seeds are considered tonic, emetic, antiperiodic and anthelmintic. The seeds, stems and bark are poisonous.

Chemistry: Two amorphous saponins, which are nearly alike in chemical and pharmacological properties, have been isolated from the seeds in a total yield of 5.5%. They have a strong haemolytic action on human red blood cells.

Erythrina stricta Roxb.

Family: Leguminosae (Fabaceae)

Medium sized, armed, diciduous tree. Leaves pinnately 3-foliolate, leaflets rhomboid-ovate, glabrescent, thin coriaceous. Flowers deep red crowded in 10 cm long racemes. Fruit pod, 3–4-seeded, to 10 cm long.

Occasional in deciduous forests.

Fl. & Fr.: January–April.

Sanskrit: Mura, Paribhadra. Telugu: Mullu Moduga

Medicinal uses: The bark in powdered form is used in biliousness, rheumatism, itch, asthma, leprosy, epilepsy. The flowers are said to be given as an antidote to poison. Crushed leaves applied to rheumatic joints helps to relieve pain.

Erythrina suberosa Roxb.

Family: Leguminosae (Fabaceae)

Deciduous prickly trees, up to 10 m tall. Leaves 3-foliolate, leaflets broadly ovate, glabrous above, tomentose below. Flowers red in 8 cm long racemes. Pod stipitate, torulose, 2–5-seeded.

Occasional in deciduous forests.

Fl. & Fr.: February–May.

Sanskrit: Mandala. Telugu: Mulumoduga.

Trade name: Corky coral tree.

Medicinal uses: The bark is reported to be used in preperation of midicines for dysentery. Leaves are reported to be used to destroy maggots in ulcers.

Erythrina variegata L.

Syn: *Erythrina indica* Lam.

Family: Leguminosae (Fabaceae).

Small armed deciduous trees. Leaves trifoliolate, leaflets rhomboid-ovate, thin, coriaceous. Flowers bright red, in 5–15 cm long racemes. Fruit 6–8-seeded pod, seeds red to dark purple or brown.

Common in waste lands, bushes and deciduous forests.

Fl. & Fr.: February–June.

Sanskrit: Paribhadra, Raktapushpa. Telugu: Badisa. Boditha.

Trade name: Indian coral Tree.

Medicinal uses: The bark is astringent and used as febrifuge, anti-biliousness and anthelmintic. It is useful as a collyrium in ophthalmia. Different parts of the plant have been used in traditional medicine as a nervine sedative, antiasthmatic and antiepileptic.

Eucalyptus globulus Labill.

Family: Myrtaceae.

Large trees, bark smooth, sheds in long strips or sheet. Leaves elliptic-ovate, lanceolate or falcate. Flowers whitish, solitary or rarely three, axillary. Fruit globular, sessile.

Planted on roadsides, wastelands and degraded forest by Forest Department.

Telugu: Neelagiri thailamu chettu, Eucalyptus.

Trade name: Eucalyptus.

Medicinal uses: Eucalyptus oil is obtained from leaves by steam distillation. The oil is used locally as an antiseptic, especially in the treatment of infections of the upper respiratory tract, and in the certain skin diseases. Mixed with an equal amount of olive oil, it is useful as a rubifacient for rheumatism. It is also used in ointments for burns. Internally it is used as a stimulating expectorant in chronic bronchits and asthma.

Euphorbia acaulis Roxb.

Syn: *Euphorbia fusiformis* Buch-Ham.

Family: Euphorbiaceae.

Perennial herbs, stem reduced to a stout underground root stock, fusiform. Leaves in apical rosettes, obovate or oblanceolate, fleshy, subsessile. Flowers in cyathia, appear after the leaves have fallen. Male flowers many, female flower solitary. Fruit capsule.

Rare in Cuddapah, Kurnool and Warangal districts.

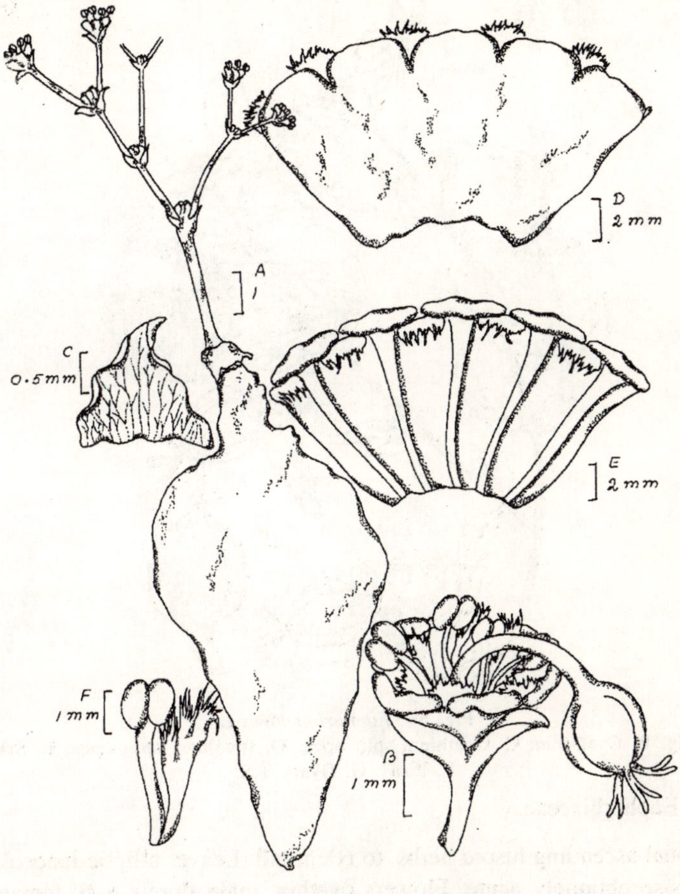

Fig. 58: *Euphorbia acaulis* Roxb.
A. Twig; B. Cyathium; C. Bract; D. Involucre split open (without); E. Involucre split open (within); F. Stamen with bracteole.

Fl. & Fr.: February–September.
Telugu: Barresepugadda.

Medicinal uses: The n-Hexane fraction of the plant extract is reported to possess potent anti-inflammatory activity. The active principle isolated is euphol, exhibiting antiinflammatory activity equivalent to phenyl butazone. Euphol is reported to possess significant hypotensive activity when administered intravenously and anti-inflammatory activity on intraperitonial injection. The rhizome contains diterpenes.

Euphorbia hirta L.

Syn: *E. pilulifera* sensu Hook.f. non Linn.

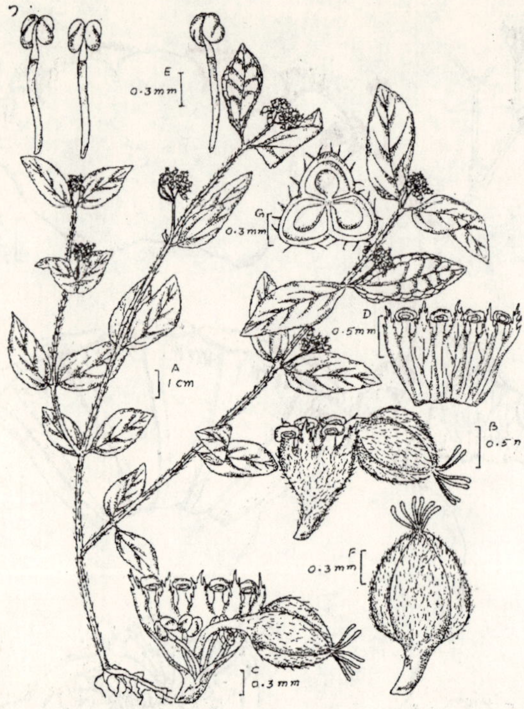

Fig. 59: *Euphorbia hirta* L.
A. Twig; B. Cyathium; C. Cyathium split open; D. Involucre split open; E. Stamens; F. Pistil; G. Ovary T.S.

Family: Euphorbiaceae.

Annual ascending hispid herbs, to 60cm tall. Leaves elliptic-lanceolate, opposite, base obliquely acute. Flowers cyathia, male florets 4–6, female floret laterally pendulous. Fruit capsule, pubescent, seeds 4-angled.

Common weed along road sides, wastelands and cultivated fields.

Fl.& Fr.: Through out the year.

Sanskrit: Pusitoa. Telugu: Reddivarinanubalu.

Trade name: Snake weed.

Medicinal uses: The entire plant, collected in flowering and fruiting stage and dried, constitutes the drug. *Euphorbia hirta* causes relaxation of bronchitis and has a depressent action on heart and respiration. It is useful in removing worms in children, in bowel complaints, asthma and cough. It promotes formation and flow of milk in women; it is also useful in gonorrhoea and other urinogenitary complaints. The roots of the plant stop vomiting.

Euphorbia nivulia Buch-Ham.

Family: Euphorbiaceae.

Succulent trees, up to 9m tall, cylindrical, with sharp spines, arranged in regular rows. Leaves obovate-oblong, deciduous. Flowers in axillary cyathia, male florets in 5 groups of eight each, female florets laterally pendulous. Fruit capsule.

Commonly along rocky areas in open dry forests.

Fl. & Fr.: November–January.

Sanskrit: Patrasuchi. Telugu: Bontha Jamudu, Aku jamudu.

Trade name: Leafy milk hedge, Holy milk hedge.

Medicinal uses: The latex of this plant possess vesicant, wormicidal and purgative properties.

Euphorbia tirucalli L.

Family: Euphorbiaceae.

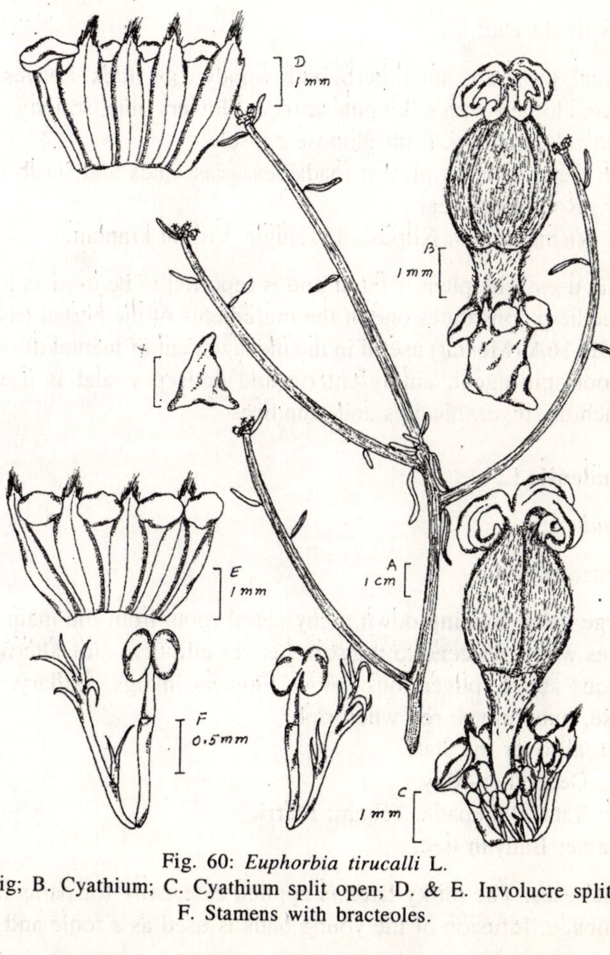

Fig. 60: *Euphorbia tirucalli* L.
A. Twig; B. Cyathium; C. Cyathium split open; D. & E. Involucre split open; F. Stamens with bracteoles.

Succulent shrub or small tree, up to 5m tall, branchlets terete, glabrous. Leaves linear, caducous, subsessile. Florets in cyathia, terminal male florets 6–8, female florets erect. Fruit globose capsule, pubescent, seeds globose.

Occasional in field hedges in dry localities.

Fl. & Fr.: January–April.

Sanskrit: Trikantaka, Vajradruma, Kadondasnuti. Telugu: Kalli, Sanna jamudu.

Trade name: Round milkhedge, Milk bush.

Medicinal uses: The white juice of the plant is considered useful for local application in rheumatic pains, toothache etc. It is also useful in neuralgia, rubifacient, cough and cold.

Evolvulus alsinoides (L.) L.

Syn: *Convolvulus alsinoides* L.

Family: Convolvulaceae.

A perennial, prostrate, small herb, with woody root stock. Leaves alternate, elliptic-oblong, clothed with silky pubescence. Flowers blue, solitary or in pairs with long pedicels, axillary. Fruit globose capsule.

Common weed of open places, roadsides, grasslands and scrub jungles.

Fl. & Fr.; Round the year.

Sanskrit: Vishnukranta, Nilpushpi. Telugu: Vishnu krantamu.

Medicinal uses: The plant is bitter and is reported to be used as a tonic and febrifuge. The herb constitutes one of the ingredients of the herbal psychotropic preperation BR-16A (Mentat) useful in the management of mental disorders. The herb is a Nootropic agent, antidysenteric and antiseptic and is used to cure chronic bronchitis, fever, hiccups and jaundice.

Ficus benghalensis L.

Syn: *Ficus indica* L.

Family: Moraceae.

Very large trees, sending down many aerial roots from the main trunk and large branches which descend to the soil. Leaves elliptic-ovate, alternate, coriaceous, glabrous above, puberulous below. Flowers in figs, axillary, sessile, in pairs, globose, puberulous, red when ripe.

Common all over the state.

Fl.& Fr.: December–May.

Sanskrit: Vata, Bahupada. Telugu: Marri.

Trade name: Banyan tree.

Medicinal uses: The milky latex is applied externally for pains in rheumatism and lumbago. Infusion of the young buds is used as a tonic and is used in

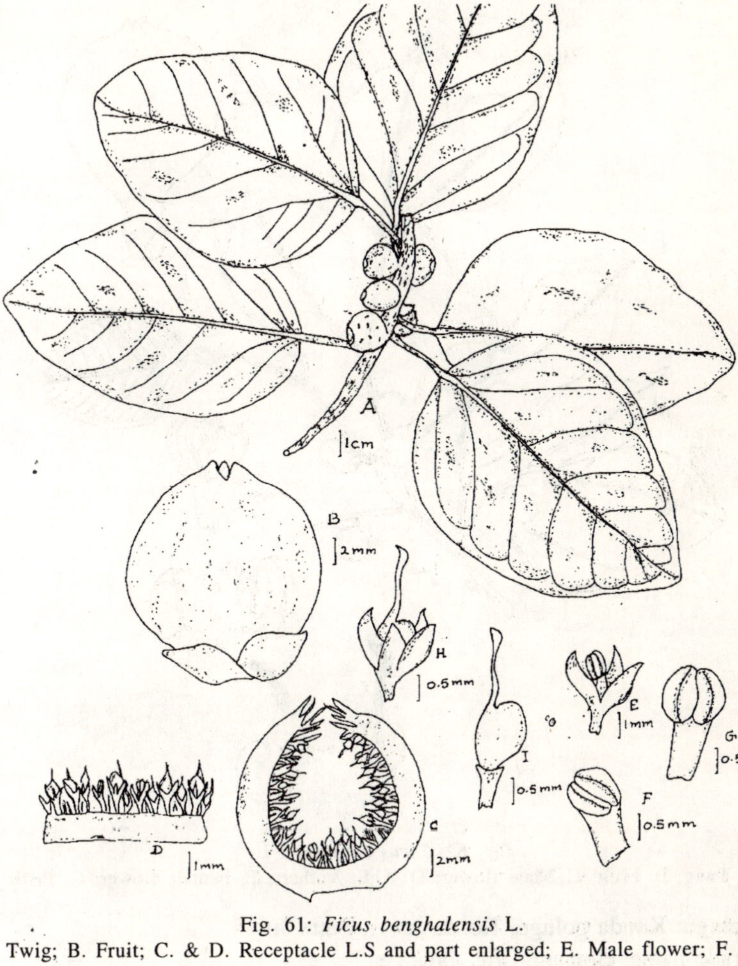

Fig. 61: *Ficus benghalensis* L.
A. Twig; B. Fruit; C. & D. Receptacle L.S and part enlarged; E. Male flower; F. & G. Anthers; H. Female flower; I. Pistil.

treating dysentery, diarrhoea, and diabetes. Seeds make a cooling tonic. Leaves are applied as poultice to abscesses.

Ficus benjamina L.

Family: Moraceae.

Large trees with numerous small aerial roots, all parts are glabrous. Leaves elliptic-ovate, alternate, subcoriaceous. Flowers in axillary figs, figs sessile, paired, globose, yellowish when ripe.

Occasional.

Fl. & Fr.: January–April

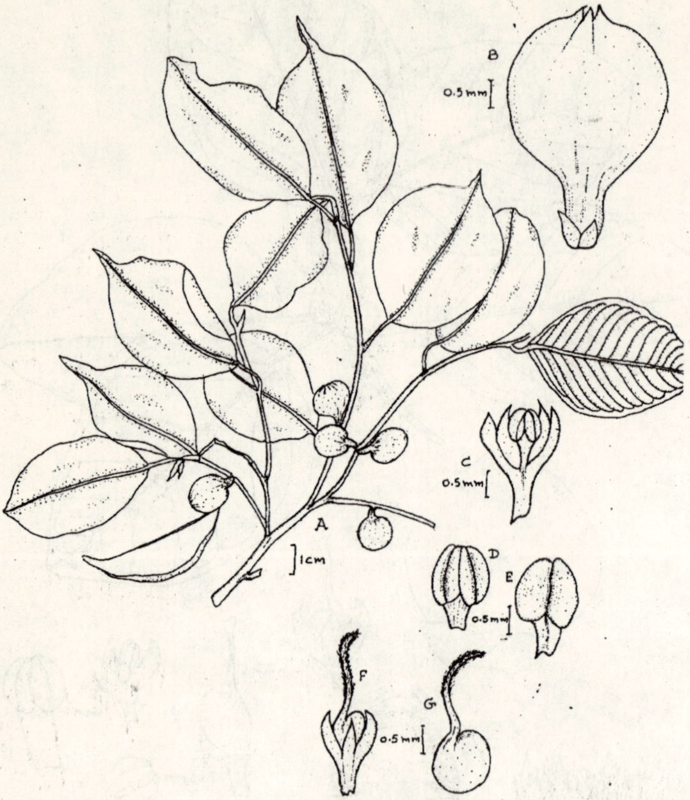

Fig. 62: *Ficus benjamina* L.
A. Twig; B. Fruit; C. Male flower; D. & E. Anthers; F. Female flower; G. Pistil.

Telugu: Konda golugu, Konda juvvi, Putra Juvvi.
Trade name: Benjamin fig, Java fig.

Medicinal uses: Leaves are used in ulcers.

Ficus carica L.

Family:Moraceae

Small or modertate sized deciduous trees. Leaves ovate or nearly orbicular, more or less deeply 3–5-lobed, rough above and pubescent below. Friut axillary, usually pear-shaped, variable in size and colour.
Cultivated occasionally.
Sanskrit: Ansira. Telugu: Anjooramu.
Trade name: Common fig.
Propagation: by stem cuttings.

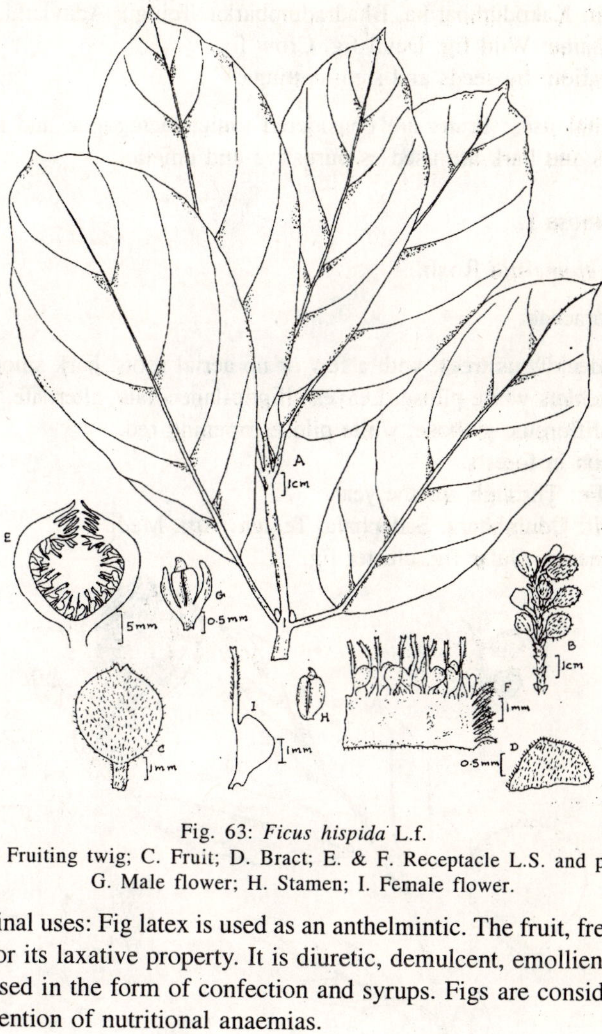

Fig. 63: *Ficus hispida* L.f.
A. Twig; B. Fruiting twig; C. Fruit; D. Bract; E. & F. Receptacle L.S. and part enlarged; G. Male flower; H. Stamen; I. Female flower.

Medicinal uses: Fig latex is used as an anthelmintic. The fruit, fresh or dried, is valued for its laxative property. It is diuretic, demulcent, emollient and nutritive. It is used in the form of confection and syrups. Figs are considered useful in the prevention of nutritional anaemias.

Ficus hispida L.f.

Family: Moraceae.

Shrub or small trees, generally with hollow internodes, without aerial roots, bark grey, rough, all parts very hispid. Leaves sub-obovate or ovate-oblong, opposite, at times alternate. Figs on special shoots and branchlets at the base, densely hispid, yellow when ripe.

Common in forests in damp localities.

Fl. & Fr.: December–March.

Sanskrit: Kakodumbarika, Bhadradumbarka. Telugu: Adaviathi, Kakimedi.
Trade name: Wild fig, Devil fig, Crow fig.
Propagation: by seeds and stem cuttings.

Medicinal uses: Fruits are considered tonic, lactogogue and emetic. The fruits, seeds and bark are used as purgative and emetic.

Ficus racemosa L.

Syn: *Ficus glomerata* Roxb.

Family: Moraceae.

Large deciduous trees, with a few or no aerial roots, bark smooth, greyish white, branchlets white-pilose. Leaves elliptic-lanceolate, alternate. Figs monoecious, cauliflorous, globose, white pilose, ripening red.
Common in forests.
Fl. & Fr.: Through out the year.
Sanskrit: Udumbhara, Sadaphala. Telugu: Atti, Medi.
Trade name: Gular fig, cluster fig.

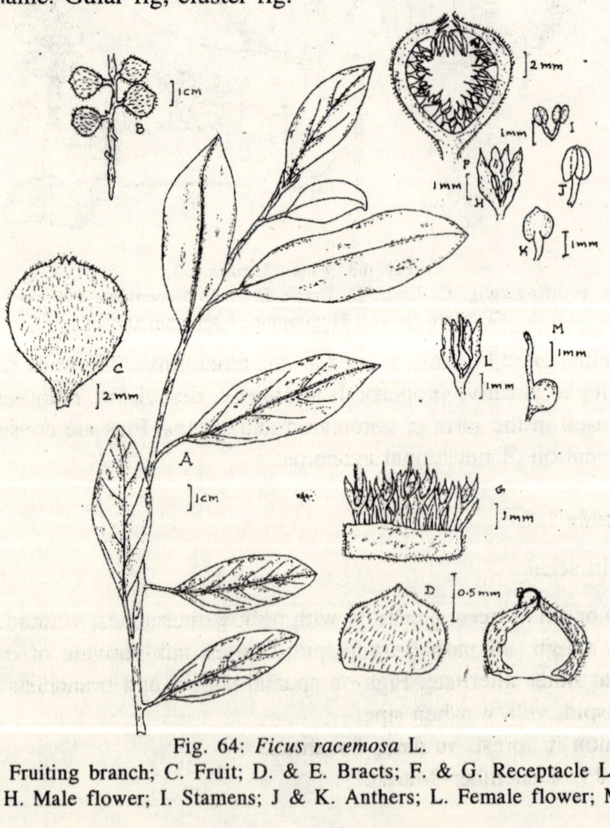

Fig. 64: *Ficus racemosa* L.
A. Twig; B. Fruiting branch; C. Fruit; D. & E. Bracts; F. & G. Receptacle L.S and part enlarged; H. Male flower; I. Stamens; J & K. Anthers; L. Female flower; M. Pistil.

Propagation: by seeds and stem cuttings.

Medicinal uses: The bark is astringent and a decoction of it is used as a wash for wounds. The root is reported to be useful in dysentery. The leaves ground to powder and mixed with honey are given in bilious affections. The fruits are astringent, stomachic and carminative. The milky latex is administered in piles and diarrhoea.

Ficus religiosa L.

Family: Moraceae.

Large glabrous trees, bark greenish. Leaves broadly ovate, alternate. Figs monoecious, axillary, sessile, paired, obovoid or depressed globose, ripening dark purple or black.

Commonly planted on village sites and as an avenue tree.

Fl. & Fr.: May–August.

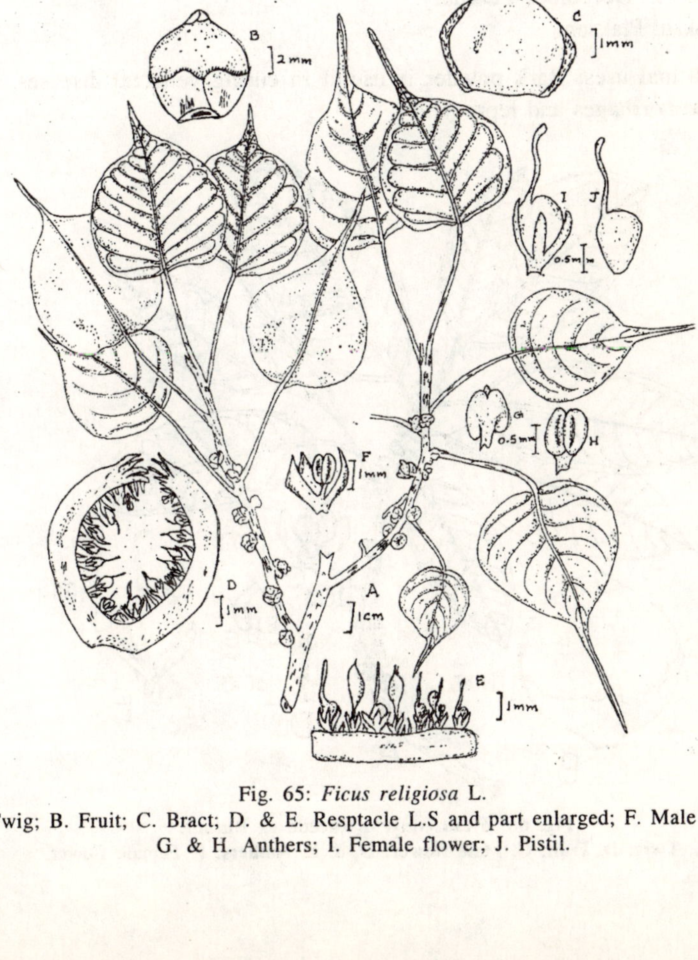

Fig. 65: *Ficus religiosa* L.

A. Twig; B. Fruit; C. Bract; D. & E. Resptacle L.S and part enlarged; F. Male flower; G. & H. Anthers; I. Female flower; J. Pistil.

Sanskrit: Ashwatha, Pippala. Telugu: Raavi.
Trade name: Peepal tree, Sacred Fig.
Propagation: by seeds.

Medicinal uses: The bark is astringent and used in the treatment of gonorrhoea, diarrhoea and dysentery. Leaves and tender shoots are used as purgative and skin diseases. The fruit is laxative and the seeds are considered to be cooling, alterative and laxative.

Ficus talboti King

Family: Moraceae.

Large evergreen trees without aerial roots, all parts glabrous, milky juice abundant, yellowish. Leaves broadly ovate-elliptic or ovate, alternate. Figs monoecious, axillary, in pairs, sessile, yellowish when ripe.

Occasional in forests.

Fl. & Fr.: December–February.

Sanskrit: Plaksah.

Medicinal uses: Bark powder is useful in curing venereal diseases, diarrhoea, haemorrhages and leprosy.

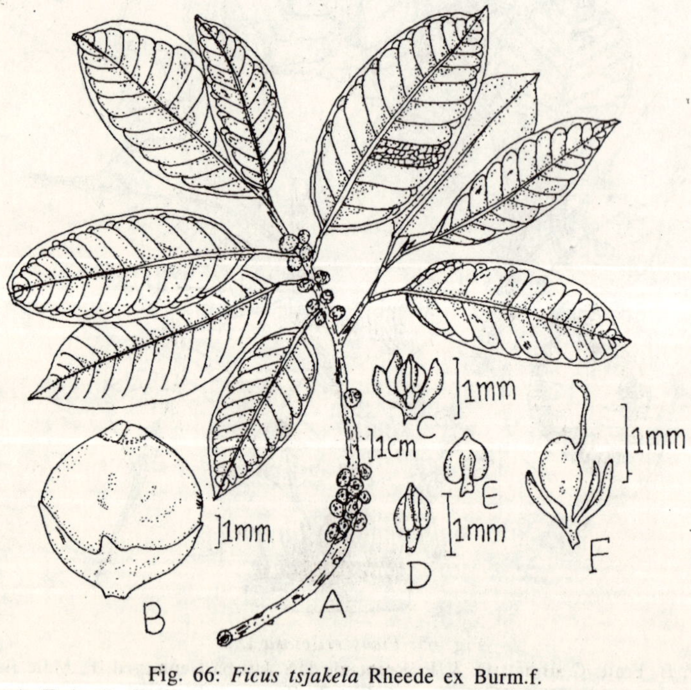

Fig. 66: *Ficus tsjakela* Rheede ex Burm.f.
A. Twig; B. Fruit; C. Male flower; D. & E. Anthers; F. Female flower.

Ficus tsjakela Rheede ex Burm.f.

Family: Moraceae.

Large glabrous trees, without aerial roots, bark rough, dark brown. Leaves broadly obovate-oblong, alternate. Figs monoecious, in clusters of 2–6 in the axils of leaves or in the axils of the scars of fallen leaves, whitish yellow, dotted when ripe.

Occasional in forests of Mahaboobnagar and Visakhapatnam districts.

Fl. & Fr.: January–May.

Sanskrit: Kapitanah, Plaksah. Telugu: Kallal, Pepre.

Medicinal uses: The bark is astringent and slightly bitter. It has been found to have antimicrobial properties and is reportedly being used to wash or clean sores.

Flacourtia indica (Burm.f.) Merr.

Syn: *Gmelina indica* Burm.f.
Flacourtia ramontchi L'Herit.
Flacourtia sepiaria Roxb.

Family: Flacourtiaceae.

Shrubs or small trees, up to 4 m tall. Leaves usually clustered at apex, basally opposite or alternate, ovate or obovate. Flowers small, white, in short racemes. Fruit globose drupe, red when ripe, 3–5-seeded.

Occasional in Scrub jungles.

Fl & Fr.: June–March.

Sanskrit: Tailsa, Aghori. Telugu: Kanregu, Pulivelaga.

Trade name: Governer's Plum, Madagascar Plum.

Propagation: by seeds.

Medicinal uses: The fruits are appetising and digestive. They are given in jaundice and enlarged spleen. The bark is considered astringent and diuretic. The root paste is applied externally in skin diseases.

Gardenia gummifera L.f.

Family: Rubiaceae.

A small tree, up to 3m tall. Leaves oblong-obovate. Flowers axillary, solitary, yellowish white or white. Fruit oblong or ellipsoid berry.

Common in deciduous forests.

Fl. & Fr.: January–September.

Sanskrit: Nadihingu, Pindava. Telugu: Bikki, Manchibikki.

Trade name: Dikamali, Cumbi gum.

Propagation: by stem cuttings.

Medicinal uses: The leaf buds and young shoots yield a resinous exudation, known in commerce as Dikamli or Cumbi gum. The resin is antispasmodic, expectorant, carminative, diaphoretic and anthelmintic. It is given to children in nervous disorders and diarrhoea due to dentition and rubbed on gums to allay irritation. It is also used to clean foul ulcers. Externally it acts as an antiseptic and stimulant.

Gendarussa vulgaris Nees

Syn: *Justicia gendarussa* Burm.f.

Family: Acanthaceae.

An erect undershrub, to 2m tall, branchlets terete. Leaves narrowly lanceolate, glabrous. Flowers white with purple lines, clustered in terminal spikes. Fruit capsule.

Cultivated, frequently run wild.

Fl. & Fr.: August–January.

Sanskrit: Nila-nirgundi. Telugu: Nallavavili, Nelavavili.

Trade name: Black purification Plant.

Propagation: by seeds.

Medicinal uses: The plant is considered febrifuge, emetic, emmenagogue and diaphoretic. The leaves are used as antiperiodic, alterative and insecticide. Leaves and tenders shoots are considered diaphoretic and an infusion of the leaves is given internally in cephalagia, hemiplegia and facial paralysis.

Givotia moluccana (L.) Sreem.

Syn: *Givotia rottleriformis* Griff.
Croton moluccanum L.

Family: Euphorbiaceae.

Trees, to 20m tall, branchlets stellately tomentose. Leaves broadly ovate, cordate, sparsely stellate tomentose above, densely woolly stellate tomentose below. Flowers in lax axillary or subterminal panicles. Fruit drupe, subglobose, fulvous tomentose, seeds globose.

Occasional in forests.

Fl. & Fr.: March–July.

Telugu: Tella poliki.

Trade name: White Catamaran Tree.

Propagation: by seeds.

Medicinal uses: Bark powder and seeds are used in rheumatism, dandruff and psoriasis.

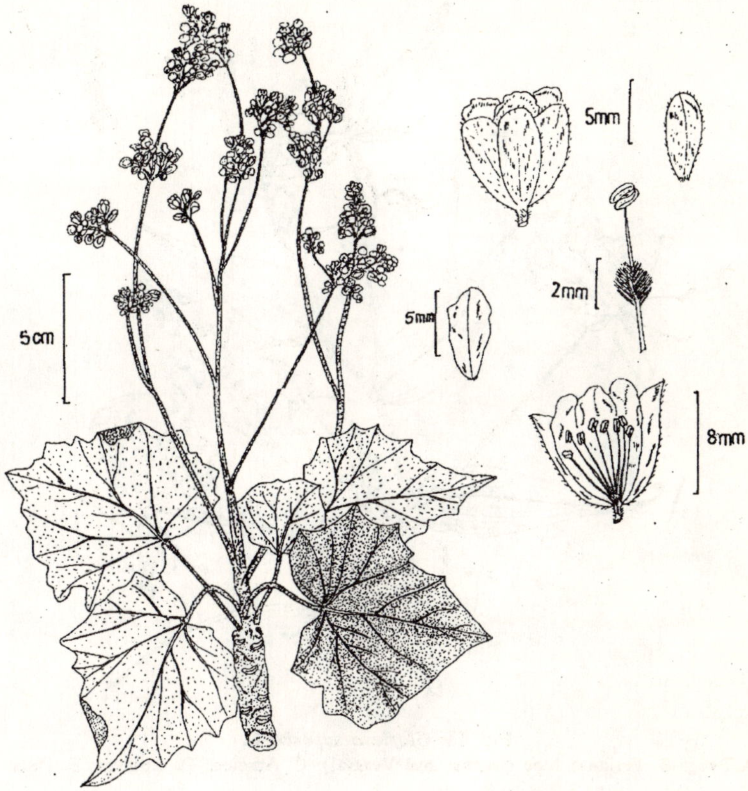

Fig. 67: *Givotia moluccana* (L.) Streem.
A. Twig; B. Flower entire; C. Sepal; D. Petal; E. Stamen; F. L.S of flower.

Gloriosa superba L.

Family: Liliaceae.

Climbing herbs, root stock with cylindric tubers, branchlets terete. Leaves alternate, opposite or whorled, oblong-lanceolate, sessile, glabrous, base cordate, margin entire. Flowers large, axillary, solitary at the ends of branches, greenish at first, then yellow, passing through orange and scarlet to crimson. Fruit oblong-ellipsoid capsule, seeds numerous, globose, warty.

Common in all districts, climbing on bushes and hedges.

Fl. & Fr.: August–September.

Sanskrit: Langali, Agnisikha. Telugu: Adavi nabhi, Kalappa gadda, Vanka vajramu.

Trade name: Glory lily.

Propagation: Through seeds collected during october and through tubers.

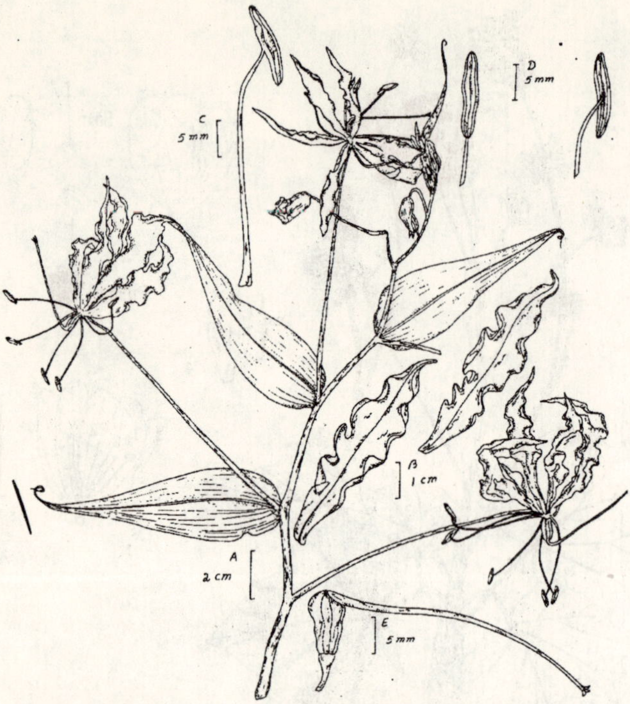

Fig. 68: *Gloriosa superba* L.
A.Twig; B. Perianth lobe (Dorsal and Ventral); C. Stamen; D. Anthers; E. Pistil.

Medicinal uses: The tubers are regarded as tonic, stomachic and anthelmintic when taken in doses of 5–10 grams; in larger, doses they are intensely poisonous. The tubers are used for a variety of medicinal purposes. It is gastrointestinal irritant and may cause vomiting and purging. It is sometimes used for promoting labour pains and also as abortifacient. It is considered useful in colic, chronic ulcers and piles. Externally it is used as a local application for parasitic skin diseases and as a cataplasm in neuralgic pains.

Chemistry: Root tubers and seeds rich source of colchicine.

Glycosmis pentaphylla (Retz.) DC.

Syn: *Limonia pentaphylla* Retz.
Glycosmis cochinchinensis auct.

Family: Rutaceae.

Shrubs. Leaves compound, 3–7-foliolate, leaflets usually paler (Whitish) beneath. Flowers in dense axillary racemes. Fruit subglobose berry, white to pink.

Occasional in forests of North coastal, Chittor and Cuddapah districts.

Fig. 69: *Glycosmis pentaphylla* (Retz.) DC.
A. Twig; B. Bract; C. Calyx; D. Corolla lobe; E. Stamens; F. Pistil.

Fl. & Fr.: June–August.
Sanskrit: Pithari. Telugu: Gogulu.
Propagation: by seeds and stem cuttings.

Medicinal uses: The plant is used in indigonous medicine for cough, rheumatism, anaemia and jaundice. The juice of the leaves, which is bitter, is used in fever, liver complaints and as vermifuge. A paste of the leaves with ginger is applied in eczema and skin affections. A decoction of the root is given for facial inflammations.

Chemistry: The leaves contain a glycoside, glycosmin.

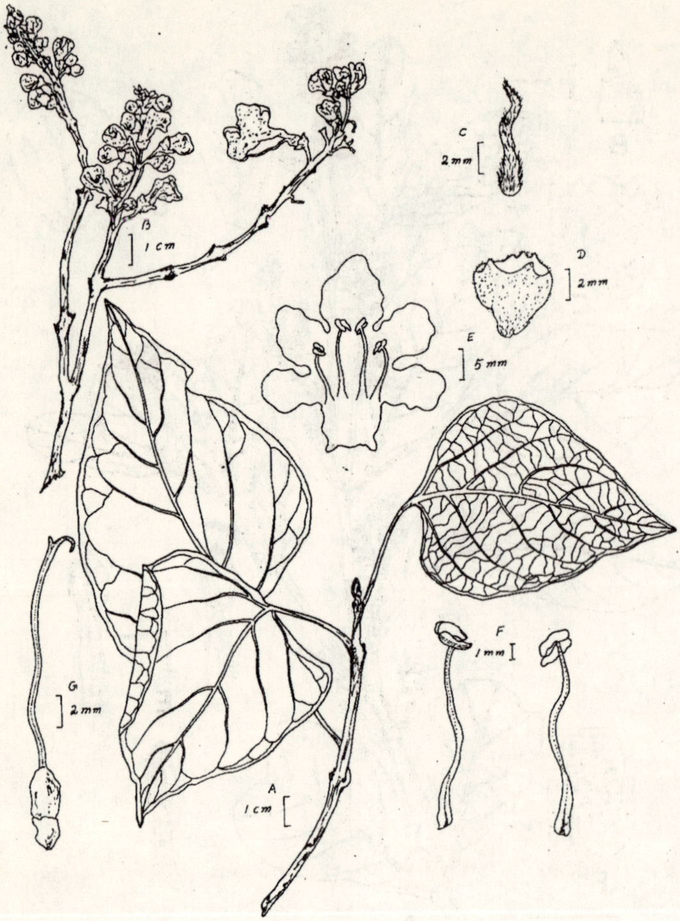

Fig. 70: *Gmelina arborea* Roxb.
A. Branch; B. Inflorescence; C. Bract; D. Calyx; E. Corolla open; F. Stamens; G. Pistil.

Gmelina arborea Roxb.

Family: Verbenaceae.

Unarmed medium sized trees, up to 10m tall, branches terete. Leaves broadly ovate. Flowers yellow or yellowish brown in threes in panicles. Fruit ovoid or pyriform drupe, orange yellow when ripe.

Occasional in the deciduous forest.

Fl.. & Fr.: March–May.

Sanskrit: Gambhari, Kasmari. Telugu: Gummadi teku.

Trade name: Coomb Teak, White Teak.

Medicinal uses: The root is an ingredient of the ayurvedic preperation "Dasamula". The root is a bitter tonic, stomachic, laxative, galactogogue. In the form of infusion or decoction it is prescribed in indigestion, fevers and anasarca. Pulverised root is applied locally for gout. The bark is a bitter tonic and stomachic, and is considered useful in fever and indigestion. The fruits, which are sweetish and bitter, are used as an ingredient of refrigerant decoctions for fevers and bilious affections. The tender leaves are demulcent. A paste of the leaves is applied to the head for relief of headache in fevers. The flowers are given in blood diseases.

Gmelia asiatica L.

Family: Verbenaceae.

Large armed straggling shrubs. Leaves elliptic-ovate. Fruit ovoid drupe, yellow when ripe.

Occasional in scrub forests.

Fl. & Fr.: May–October

Sanskrit: Biddari, Gopabhadra, Vikarini. Telugu: Adavi Gummudu, Challa Gimmudu.

Trade name: Small Cashmere tree.

Medicinal uses: Fruits are used in Gonorrhoea, Dysuria, inflammations and swellings. Fruits are also used as hair tonic and to clear dandruff. The root is aromatic, mucilaginous, demulcent, alterative and astringent. It is used in gonorrhoea, catarrah of bladder, rheumatism and as blood purifier.

Gossypium arboreum L.

Family: Malvaceae.

Shrubs, to 3m tall, some times arborescent. Leaves palmately 3–5-lobed or parted, the lobes lanceolate. Flowers pale yellow, usually with a purplish centre, axillary, solitary. Fruit loculicidal capsule, ovoid, seeds many with dense white hairs.

Extensively cultivated for cotton.

Fl. & Fr.: August–April.

Sanskrit: Raksatika. Telugu: Pratti.

Trade name: Cotton.

Propagation: by seeds.

Medicinal uses: Seeds are considered to be useful against rheumatism where as the leaves are diuretic and are useful in the treatment of skin diseases.

Gymnema sylvestre (Retz.) R. Br. ex Schult.

Syn: *Periploca sylvestris* Retz.

Family: Asclepiadaceae.

Large, woody much branched climber running over the tops of high trees, young stems and branches pubescent. Leaves ovate-elliptic or ovate-lanceolate. Flowers minute, greenish-yellow or tinged pink in axillary pedunculate cymes. Fruit follicles, beaked at apex, glabrous. Seeds with thin broad marginal wing.

Occasional in forests in all districts.

Fl. & Fr.: September–December.

Sanskrit: Madhunasini, Mesharingi. Telugu: Podapatri.

Trade name: Periploca of the woods, sugar killer.

Propagation: by stem cuttings and by seeds.

Medicinal uses: Leaves used for treating diabetes. Cordiac stimulant. The plant is stomachic, stimulant, laxative and diuretic. It is said to be useful in cough, biliousness and sore eyes. The leaves of the plant, when chewed, possess the remarkable property of paralysing, for a few hours, the sense of taste for sweet and bitter substances. The plant is bitter astringent and useful in inflammations, dyspepsia, constipation, jaundice, haemorrhoids, strangury, renal and vesical calculi, helminthiasis, cardiopathy, cough, asthma, bronchitis, intermittent fever, amenorrhoea, conjunctivitis and leucoderma. Roots are emetic and expectorant.

Haldinia cordifolia (Roxb.) Ridsd.

Syn: *Adina cordifolia* (Roxb.) Hook.f. ex Brandis
Nauclea cordifolia Roxb.

Family: Rubiaceae.

Large deciduous tree, up to 18m tall. Bark brownish-grey, thick furrowed. Leaves thick chartaceous, orbicular-ovate. Flowers yellowish in axillary heads. Fruit capsule, cuneate, densely downy, breaking into 2 follicular cocci. Seeds 6, elongate, tailed above or winged.

Common in dry deciduous forest all over Andhra pradesh.

Fl.: June–September., Fr.: February–May of the following year.

Telugu: Rudra Ganapa, Bandaru, Pasupu ganapa.

Trade name: Haldu; Turmeric wood.

Medicinal uses: The roots are used as astringent in dysentery. The bark is acrid and bitter and is used in biliousness. It is also regarded as febrifuge and antiseptic.

Hedychium coronarium Koenig

Family: Zingiberaceae

A tuberous herb, root stock horizontal, fleshy, jointed. Leaves distichous, lanceolate or oblong-elliptic, glabrous. Flowers white in spikes. Fruit oblong capsule, many-seeded.

In moist localities in north Coastal districts.

Fl. & Fr.: July–August.

Propagation: by rhizome cuttings.

Medicinal uses: A paste made from the rhizomes is reported to be applied to bruises and sprains and used as a carminative and stimulant.

Hedyotis corymbosa (L.) Lam.

Syn: *Oldenlandia corymbosa* L.

Family: Rubiaceae.

Annual erect, scaberulous herbs. Leaves linear-lanceolate, scabrid. Flowers white in axillary peduncled cymes. Fruit loculicidal capsule.

Common in forests and in cultivated fields.

Fl. & Fr.: June–January.

Sanskrit: Parpatah, Parpatakah; Rajana.

Telugu: Verri-nelavemu.

Propagation: by seeds.

Medicinal uses: The plant is considered to possess cooling, febrifugal, pectoral and stomachic properties. It is usually administered in the form of decoction in remittent fever with gastric irritability and nervous depression caused by deranged bile. It is also given in jaundice and diseases of the liver and as a cure for heat eruptions. The plant is also used as anthelmintic.

Helicteres isora L.

Family: Sterculiaceae.

Shrubs or trees, up to 5m tall, branchlets apically stellate-tomentose. Leaves obovate or elliptic-oblong, scabrous above, pubescent beneath. Flowers crimson in axillary cymes. Fruit of 5 spirally twisted follicles, seeds minutely tuberculate.

Common in forests in all districts.

Fl. & Fr.: July–December.

Sanskrit: Mrigashringa, Murva, Avarttani. Telugu: Godratada, Valumbar, Peddashamala.

Trade name: East Indian Screw Tree

Propagation: both by seeds and vegetative means.

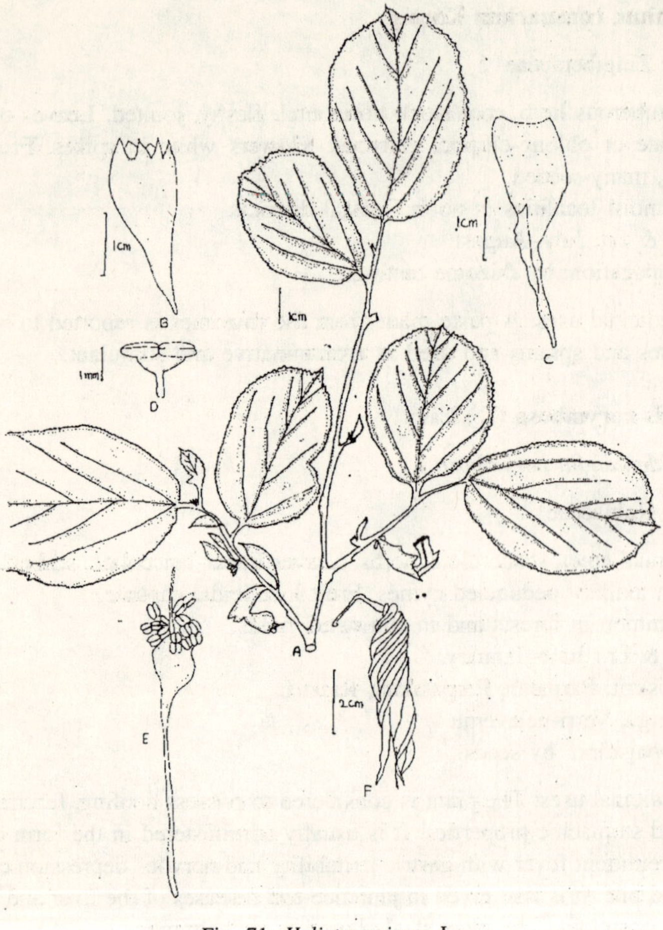

Fig. 71: *Helicteres isora* L.
A. Twig.

Medicinal uses: Dried fruit is reported to be useful in intestinal complaints and prescribed in indigenous systems for colic, flatulence ophthalmia, kermosis, wounds, ulcers, haemorrhages and diarrhoea. The root and stem bark are considered to be expectorant, demulcent, astringent and antigalactogogue. The bark is prescribed for diarrhoea, dysentery and biliousness.

Heliotropium indicum L.

Family: Boraginaceae.

Annual, erect, hispid herbs, up to 50cm tall. Leaves alternate or suboppsite, elliptic-ovate. Flowers pale violet, in helicoid spikes, extra axillary. Fruit of 2 nutlets.

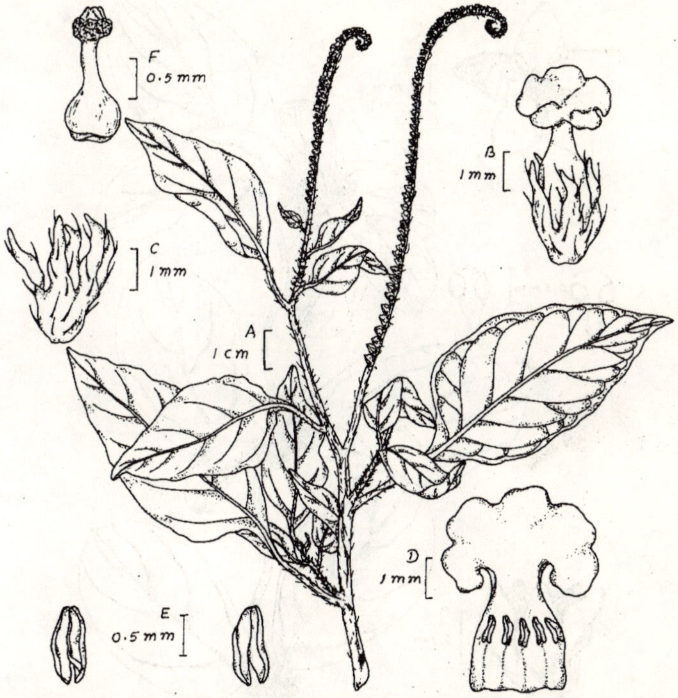

Fig. 72: *Heliotropium indicum* L.
A. Twig; B. Flower; C. Calyx; D. Corolla split open with stamens; E. Stamens; F. Pistil.

Occasional in plains, in waste lands and roadsides.
Fl. & Fr.: Round the year.
Sanskrit: Vrikshakali, Kastisunda. Telugu: Naga Danthi.
Trade name: Indian Turnsole.
Propagation: by seeds and stem cuttings.

Medicinal uses: Whole plant useful in Hydrophobia, wounds, sores, boils, opthalmia, scorpion stings, asthma, rheumatism, constipation, chest affections and leprosy. It is reported to possess emollient, vulnerary and diuretic properties. It is used as local application for ulcers, sores, wounds, gum boils, skin affections, stings of insects and rheumatism.

Hemidesmus indicus (L.) R.Br.

Syn: *Periploca indica* L.

Family: Asclepiadaceae

A perennial twining or prostrate shrub, stems numerous with greenish brown bark, root stock woody. Leaves vary greatly in shape and size, elliptic oblong to

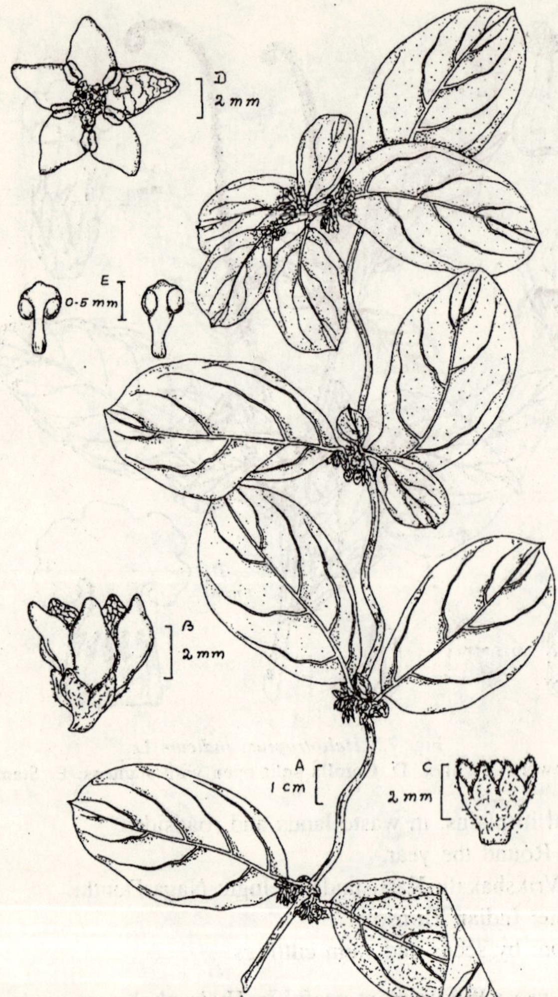

Fig. 73: *Hemidesmus indicus* (L.) R. Br.
A. Twig; B. Flower; C. Calyx; D. Flower top view; E. Pollinia.

linear-lanceolate, dark green often with whitish blotches, pubescent beneath. Flowers very small, yellowish brown, crowded in subsessile cymes in opposite axils. Fruit paired follicles, green, cylindric, pointed at tip. Seeds small, black, with a tuft of white hairs at top.

Occasional in forests and plains in all districts.

Fl.& Fr.: September–February.

Sanskrit: Anantamul, Sariva, Gopi.

Telugu: Sugandhipala, Sugandhi

Trade name: Indian Sarsaparilla.

Propagation: propagated by seeds as well as vegetative means by stem cuttings.

Medicinal uses: The dried roots are medicinal and constitute the Hemidesmus or Anantamul, which is official in Indian pharmcopoeia. Air dried roots are medicinal and used in traditional Ayurvedic medicine as a tonic, alterative, diaphoretic, blood purifier, in rheumatism and in skin infections. It is an important ingredient in more than 60 Ayurvedic preperations. It is one of the ingredient in 'Pinda Taila' an ayurvedic drug. The roots are also used in nutritional disorders, syphilis, urinary disorders, skin burning, fissures in the feet and toothache. The roots stimulate lactation. Syrup from roots is used as a flavouring agent.

Hibiscus aculeatus Roxb.

Syn: *Hibiscus furcatus* Roxb.
Hibiscus hispidissimus Griff.

Family: Malvaceae.

Rambling shrubs, stems, petioles and pedicels armed with pustular based aculei. Leaves entire to 3–5-angled or lobed. Flowers yellow with deep purple centre, axillary, solitary. Fruit capsule, ovoid or conical, seeds many, dark brown.

Rare in forests.

Fl. & Fr.: July–December.

Telugu: Adavi gogu, Konda gogu.

Propagation: by seeds.

Medicinal uses: Leaves are reported to improve digestion and are anthelmintic. The juice of leaves mixed with honey is used in the treatment of eye diseases while the roots are used in the treatment of kidney disorders, burning sensation and swellings.

Hibiscus rosa-sinensis L.

Family: Malvaceae.

A shrub, up to 3m tall. Leaves ovate, glabrous. Flowers solitary, axillary, crimson, never sets fruit.

Commonly planted as an ornamental in house gardens.

Fl. & Fr.: Round the year.

Propagation: by stem cuttings.

Sanskrit: Japapushpa, Aruna. Telugu: Mandaramu.

Trade name: Shoe flower plant, chinese rose.

Medicinal uses: The flowers and leaves are used against a variety of ailments such as alepacia, burning sensation in the body, diabetes, menstrual disorders and piles. The flowers are considered demulcent, emollient, refrigerant, aphrodisiac and emmenagogue. They are made into a paste and applied to swellings. Decoction of flowers given in bronchial catarrh. Leaves emollient and aperient. Fresh root juice is given for gonorrhoea and powdered root for menorrhagia

Hibiscus sabdariffa L.

Family: Malvaceae.

Annual erect herb, up to 1m tall, stems reddish, glabrous. Lower leaves ovate, unlobed, the cauline leaves 3-lobed, lobes crenate. Flowers pink with dark centre, axillary, solitary. Fruit ovoid capsule, pubescent, beaked.

Cultivated for its leaves (vegetable) and fibres from its stem.

Fl. & Fr.: Round the year.

Telugu: Yerra gogu.

Trade name: Rozelle.

Propagation: by seeds.

Medicinal uses: Leaves are diuretic, choretic, febrifuge and hypotensive. They are used in Dysuria, Debility, Stranguary, Dyspepsia. Preperations of the plant are used to treat carbucles, swelling and inflammation of the skin, scalding and conjunctivitis.

Active principle: Fruit acids, anthocyane, Flavonoids and mucilages.

Holarrhena pubescens (Buch.-Ham.) Wall. ex G. Don

Syn: *Holarrhena antidysenterica* Wall. ex DC.
Echites pubescens Buch.-Ham.

Family: Apocynaceae

Small tree, up to 6m tall, with milky latex, elliptic, glabrous, very large, nerves on the leaves conspicuous. Flowers white in axillary and terminal cymes, fragrant. Fruit paired follicles. Seeds with a tuft of brown hairs.

Common in forests of most districts.

Fl.& Fr.: April–October.

Sanskrit: Kutaja, Indra. Telugu: Pala kodisa

Trade name: Kurchi, Tellicherry Bark.

Propagation: By seeds.

Medicinal uses: The dried bark of the plant is used to cure amoebic dysentery. The bark has astringent, antidysenteric, anthelmintic, stomachic, febrifugal and tonic properties. It is used in the treatment of Amoebic dysentery and diarrhoea and is usually administered as extract or decoction. It's bark is given either alone or with other astringent drugs in piles, colic, dyspepsia, chest affections

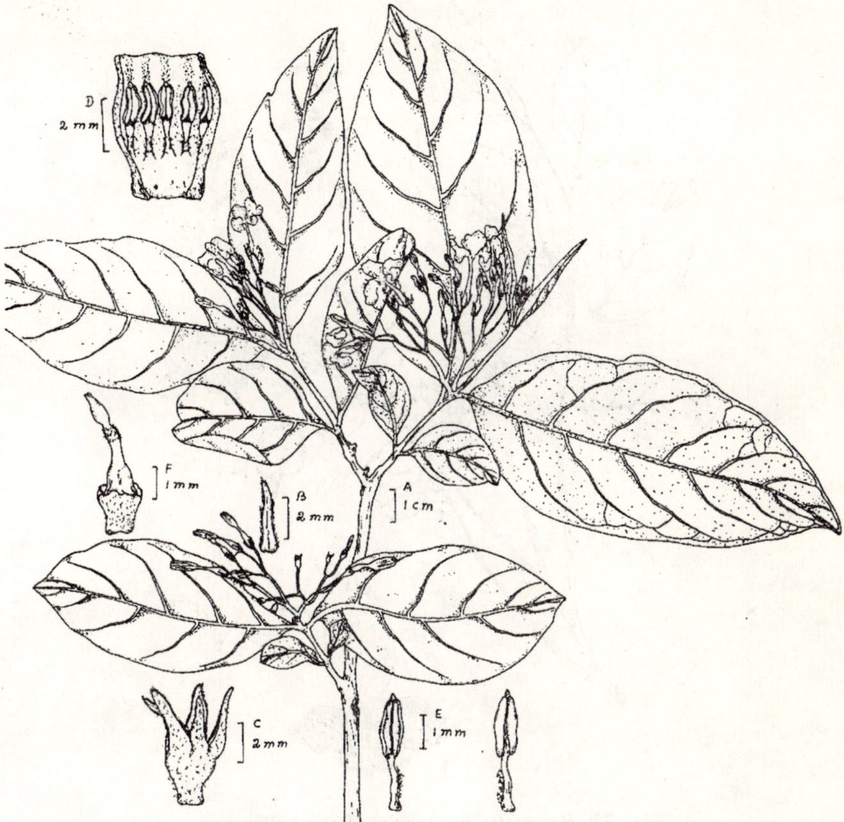

Fig. 74: *Holarrhena pubescens* (Buch-Ham.) G.Don
A. Twig; B. Bract; C. Calyx; D. Corolla split open with stamens; E. Stamens; F. Pistil.

and diuresis. It is also used for diseases of skin and spleen.

Active Principle: The alkaloid conessine present in the bark has been found to retard growth of tubercular bacilli and it is also anti-dysenteric.

Holoptelia integrifolia (Roxb.) Planch.

Syn: *Ulmus integrifolia*

Family: Ulmaceae.

Large, spreading, glabrous, deciduous trees, bark grey. Leaves ovate-elliptic, alternate, glabrous. Flowers in fascicles at the scars of old shoots, polygamous or bisexual. Fruit a dry winged compressed samara, nearly orbicular, wings membranous.

Common in forests.

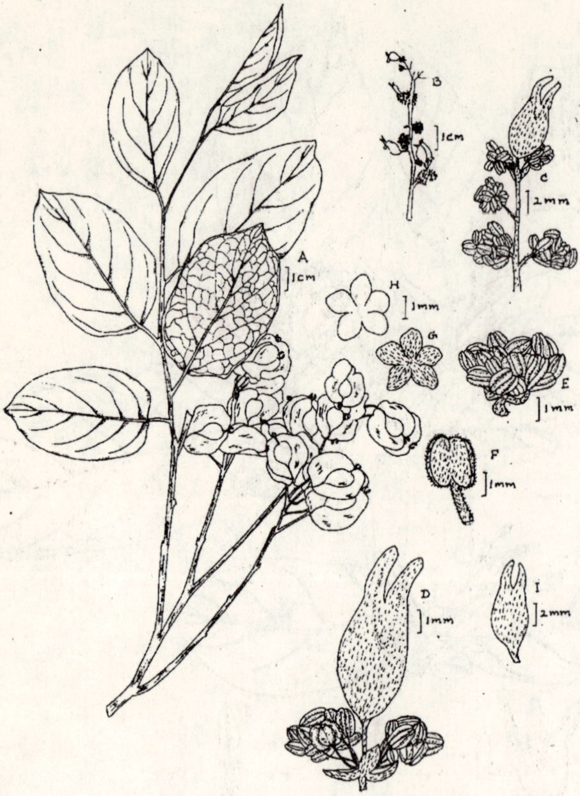

Fig. 75: *Holoptelia integrifolia* (Roxb.) Planchon
A. Twig; B. Branchlet with flowers; C. & D. Bisexual flowers; E. Male flowers; F. anthers; G. & H. Perianth; I. Ovary.

Fl.& Fr.: September–February.
Sanskrit: Cirabilvah, Putikaranjah. Telugu: Nemalinara, Pedda nevili.
Trade name: Indian elm, Kanju.
Propagation: by seeds.

Medicinal uses: The bark and leaves are used for treating inflammations, leprosy, dyspepsia, Flatulence, colic, rheumatism, helminthiasis, diabetes, skin diseases and haemorrhoids.

Holostemma ada-kodien Schult.

Syn: *Holostemma annulare* (Roxb.) K. Schum.
Asclepias annularis Roxb.

Family: Asclepiadaceae.

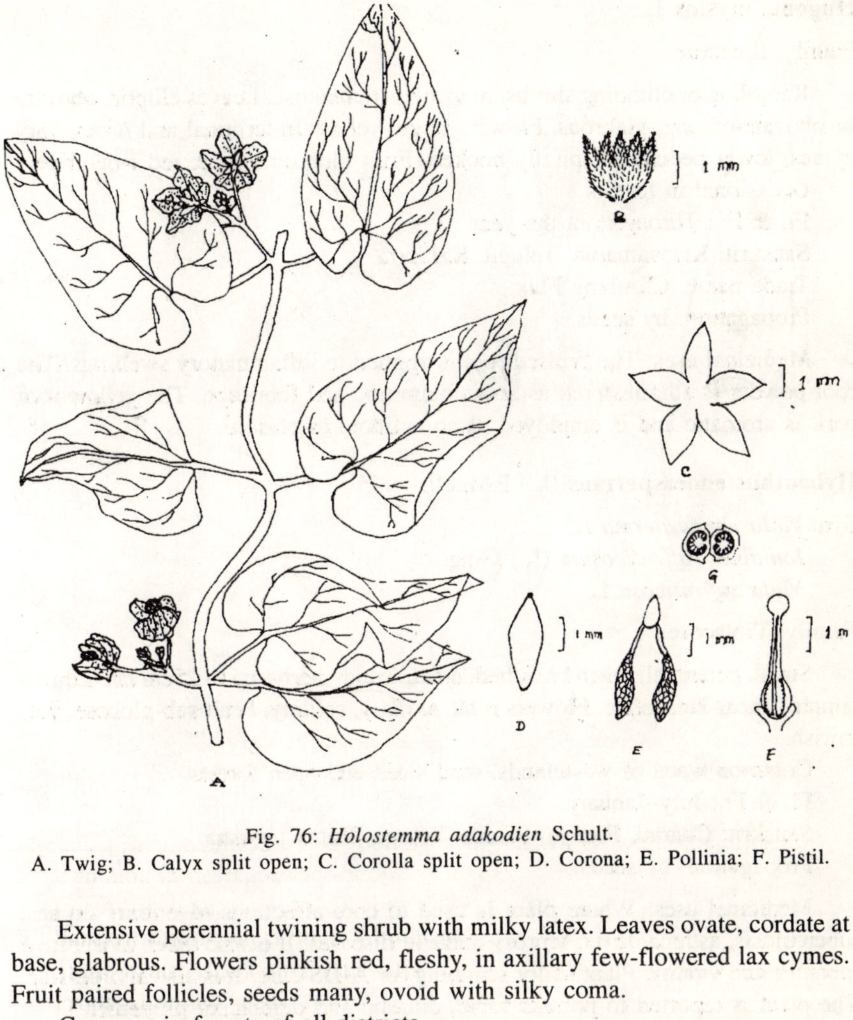

Fig. 76: *Holostemma adakodien* Schult.
A. Twig; B. Calyx split open; C. Corolla split open; D. Corona; E. Pollinia; F. Pistil.

Extensive perennial twining shrub with milky latex. Leaves ovate, cordate at base, glabrous. Flowers pinkish red, fleshy, in axillary few-flowered lax cymes. Fruit paired follicles, seeds many, ovoid with silky coma.

Common in forests of all districts.

Fl. & Fr.: April–October.

Sanskrit: Jivanti, Shitala. Telugu: Palagurugu, Dudidapalatiga, Bandiguruvinda teega.

Trade name: Ring-coronel, Swallow-wort.

Propagation: by seeds and by vegetative methods.

Medicinal uses: The roots are reported to possess cooling, alterative, tonic and lactative properties. Made into a paste they are applied in opthalmia and orchitis. They are also used in diabetes, gonorrhoea, coughs and stomach-ache.

Hugonia mystax L.

Family: Linaceae

Rambling or climbing shrubs, branchlets tomentose. Leaves elliptic, obovate or obovate-oblong, glabrous. Flowers large, yellow in terminal and /or axillary cymes, lower peduncles spirally hooked. Fruit globose, drupe, red when ripe.

Occasional in forests.

Fl. & Fr.: Through out the year.

Sanskrit: Kamsamarah. Telugu: Kakibira.

Trade name: Climbing Flax.

Propagation: by seeds.

Medicinal uses: The bruised root is applied to inflammatory swellings. The root powder is administered as an anthelmintic and febrifuge. The yellow root bark is aromatic and is employed as an antidote to poisons.

Hybanthus enneaspermus (L.) F.Muell.

Syn: *Viola enneasperma* L.
Ionidium suffruticosum (L.) Ging.
Viola suffruticosa L.

Family: Violaceae.

Small, perennial, much branched, erect, woody herbs, up to 25cm tall. Leaves simple, linear-lanceolate. Flowers pink, axillary, solitary. Fruit sub-globose, yellowish.

Common weed of wastelands, road sides and open forests.

Fl. & Fr.: July–January.

Sanskrit: Charati, Ramya, Sarada. Telugu: Ratna purusha.

Propagation: by seeds.

Medicinal uses: Whole plant is used to cure affections of urinary organs, tuberculosis, asthma, fever, leprosy and eye diseases. It is also taken to improve memory and vitality. Plant fit for scanning for AIDS cure. It is also aphrodisiac. The plant is reported to possess tonic, diuretic and demulcent properties.

Hygrophila auriculata (Schum.) Heine

Syn: *Asteracantha longifolia* (L.) Nees
Barleria auriculata Schum.
Barleria longifolia L.

Family: Acanthaceae

A stout, erect, armed unbranched strigose-hispid herb, up to 75 cm tall. Leaves whorled subsessile, 6 at a node, each having a sharp, yellow spine in its axil. Flowers bluish purple to pink, in axillary whorls. Fruit linear capsule, seeds orbicular.

Common in wet places, marshes and margins of tanks all over the state.
Fl. & Fr.: October–April.
Sanskrit: Kokilaksha. Telugu: Nirugubbi, Gobbichettu.
Trade name: Talamakhana.
Fl. & Fr.: October–April.

Medicinal uses: It is useful in dropsy, jaundice, rheumatism and diseases of urinogenital tract. Seeds useful in seminal debility and venereal diseases. The roots, leaves and seeds have been used in Indian system of medicine as diuretics, and also employed to cure jaundice, dropsy, rheumatism and ansarca.

Hyptis suaveolens (L.) Poit.

Syn: *Ballota suaveolens* L.

Family: Lamiaceae (Labiatae).

Erect, much branched aromatic undershrubs, up to 2m tall, stem tetragonal. Leaves ovate. Flowers blue or violet, scented, in axillary cymes. Fruit of nutlets.
Common in waste lands and road sides.
Fl. & Fr.: August–February.
Telugu: Sima Tulasi, Mahavira.
Propagation: by seeds.

Medicinal uses: It is considered to be stimulant, carminative, sudorific and lactagogue. An infusion of the plant is used in catarrhal conditions, affections of uterus and parasital cutaneous diseases. Leaves are used for healing wounds.

Ichnocarpus frutescens (L.) R.Br.

Syn: *Apocynum frutescens* L.

Family: Apocynaceae

Straggling shrubs, up to 10m long. Leaves oblong or elliptic-obovate, coriaceous. Flowers greenish white, numerous, in axillary and terminal, trichotomous cymes. Fruit paired follicular mericarps.
Common in forests of all districts. Fl.& Fr,: November–February.
Sanskrit: Sariva, Shymalata. Telugu: Nallatige, Karampala.
Trade name: Kaalisar.
Propagation: by vegetative means through cuttings and through suckers.

Medicinal uses: Roots are reported to be used to treat burning sensation, fever, seminal weakness, skin diseases, leprosy, vomiting, diabetes and general weakness. Powdered roots are used as a treatment for diabets and stones in the bladder. The root is also used against scorpion bites, snake bites, sores and ulcers on tongue. It is also used as a blood purifier.

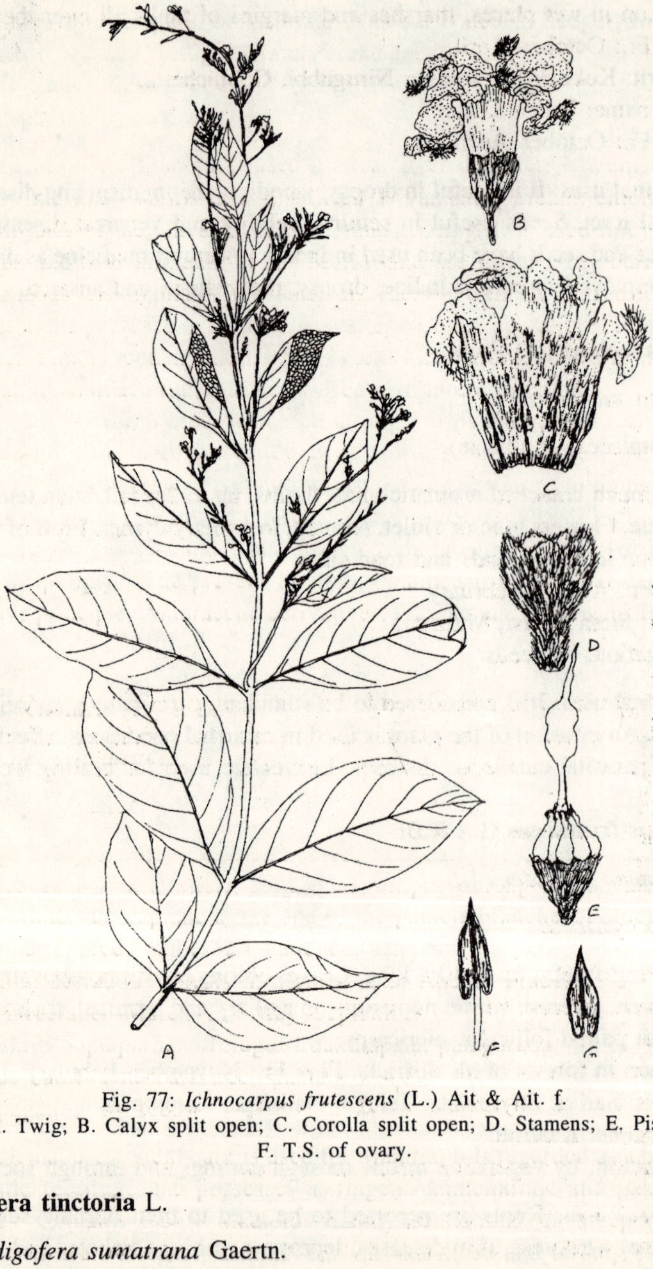

Fig. 77: *Ichnocarpus frutescens* (L.) Ait & Ait. f.
A. Twig; B. Calyx split open; C. Corolla split open; D. Stamens; E. Pistil; F. T.S. of ovary.

Indigofera tinctoria L.

Syn: *Indigofera sumatrana* Gaertn.

Family: Leguminosae (Fabaceae).

Woody, stiff undershrubs, up to 2m tall, branches terete. Leaves compound 9-foliolate, leaflets bluish-green, appressedly pubescent, ovate-oblong. Flowers

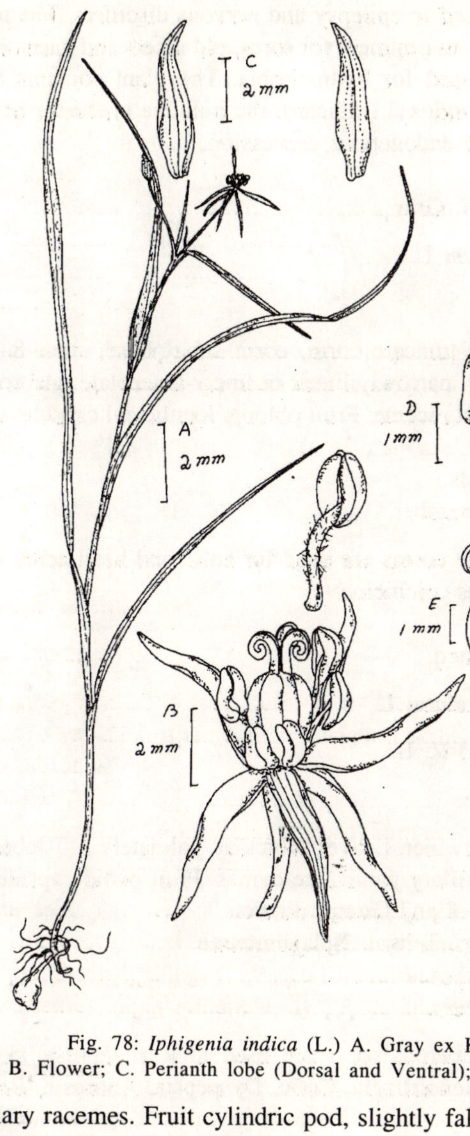

Fig. 78: *Iphigenia indica* (L.) A. Gray ex Kunth
A. Habit; B. Flower; C. Perianth lobe (Dorsal and Ventral); D. Stamen; E. Pistil.

pink in axillary racemes. Fruit cylindric pod, slightly falcate towards tip, seeds 10–12, globose.

Common in waste lands all over Andhra pradesh.

Fl.& Fr.: August–November.

Sanskrit: Nilini, Nilika. Telugu: Nili.

Trade name: Common Indigo.

Propagation: by seeds.

Medicinal uses: Whole plant given in epilepsy and nervous disorder. An extract of the plant is used in epilepsy and nervous disorders. The plant is used also in bronchitis and as an ointment for sores, old ulcers and haemorrhoids. The juice of the leaves is used for hydrophobia. The plant contains appreciable amounts of conjugated indoxyl (indican), the putative precursor of isatin. It is used in the treatment of endogenous depression.

Iphiginea indica (L.) A. Gray

Syn: *Melandrium indicum* L.

Family: Liliaceae

A herb with small tunicate corm, corm subglobose, stem 8–15 cm tall. Leaves alternate, sessile, narrowly linear or linear-lanceolate, glabrous. Flowers purple in a terminal erect raceme. Fruit oblong, loculicidal capsule, seeds ovoid, warty, brown.

Occasional in forests.
Fl. & Fr.: July–September.

Medicinal uses: The corms are used for colic and head ache.
Chemistry: Contains colchicine.

Ipomoea mauritiana Jacq.

Syn: *Convolvulus paniculatus* L.

Ipomoea paniculata (L.) R. Br.

Family: Convolvulaceae.

A tuberous woody twiner. Leaves orbicular, palmately 5–7-lobed, glabrous. Flowers rose pink in axillary paniculate cymes. Fruit ovoid capsule, glabrous.

In forests in Kurnool and Guntur districts.
Sanskrit: Ksiravidari. Telugu: Nelagummudu.
Trade name: Gaint potato
Propagation: by tubers.

Medicinal uses: Tuberous roots are used to treat leprosy, skin diseases, syphilis Gonorrhoea, Menorrhagia, Colic, Dyspepsia, Anorexia, Bronchits and fever.

Ipomoea nil (L.) Roth

Syn: *Ipomoea hederacea* auct.
Convolvulus nil L.

Family: Convolvulaceae.

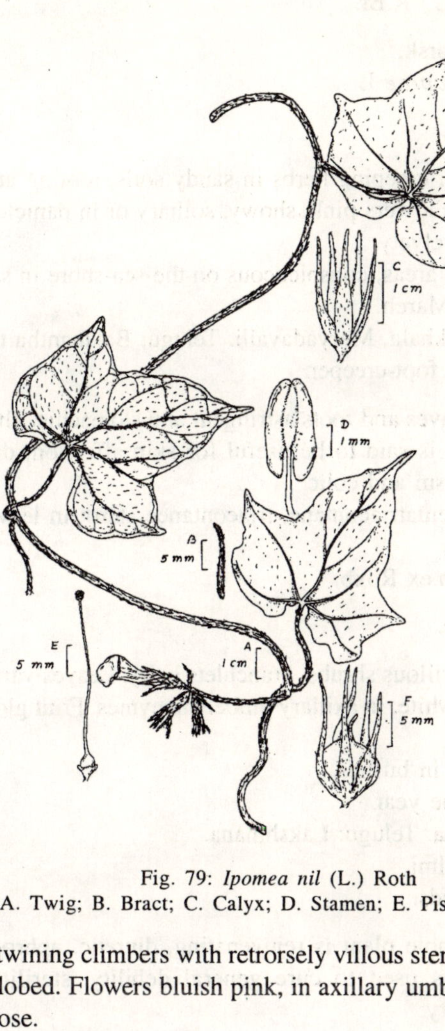

Fig. 79: *Ipomea nil* (L.) Roth
A. Twig; B. Bract; C. Calyx; D. Stamen; E. Pistil; F. Fruit.

Slender, twining climbers with retrorsely villous stem. Leaves ovate-cordate, irregularly 3-lobed. Flowers bluish pink, in axillary umbellate cymes. Fruit capsule, subglobose.

Common in jungles.

Fl. & Fr.: Most part of the year.

Sanskrit: Krisnabijah. Telugu: Kolli-vittulu.

Trade name: Morning Glory, Pharbitis seeds.

Propagation: by seeds.

Medicinal uses: The seeds are used as purgative. Both seeds and resin extract are employed as cathartic.

Ipomoea pes-caprae (L.) R.Br.

Syn: *Ipomoea biloba* Forsk.
Convolvulus pes-caprae L.

Family: Convovulaceae.

Perennial, prostrate, creeping herbs in sandy soils, rooting at nodes. Latex milky. Leaves bilobed. Flowers pink, showy, solitary or in panicles. Fruit ovoid capsule, seeds brown, velvety.

Common in coastal areas, conspicucous on the sea-shore in sandy beaches.

Fl.& Fr.: October–March.

Sanskrit: Sagaramekhala, Maryadavalli. Telugu: Balabantha theega.

Trade name: Goats foot-creeper.

Medicinal uses: Leaves and roots astringent and stomachic, alterative, tonic, diuretic and laxative. It is said to be useful for skin affection, dropsy, Gonorrhoea, phlegm rheumatism and colic.

Active principle: Pentatriacontane, triacontanee sterol in leaves and roots.

Ipomoea sepiaria Koen.ex Roxb.

Family: Convolvulaceae.

Perennial, twining villous shrubs, branchlets hairy. Leaves variable, triangular-cordiform. Flowers white, in axillary umbellate cymes. Fruit globose capsule, seeds white velvety.

Occasional climber in bushes.

Fl. & Fr.: Round the year.

Sanskrit: Lakshmana. Telugu: Lakshmana.

Trade name: Bankalmi.

Propagation: by seeds.

Medicinal uses: Whole plant is rejuvenating, diuretic, aphrodisiac, de-obstruent, uterotonic. It is used to cure general debility, sterility in women, hyperdipsia and strangury.

Ixora coccinea L.

Family: Rubiaceae.

Shrub. Leaves elliptic-obovate to oblanceolate. Flowers red or yellowish in terminal corymbose panicles. Fruit subglobose, fleshy, red in colour, crowned with persistent calyx.

Cultivated as an ornamental.

Fl. & Fr.: Round the year.

Sanskrit: Bandhuka, Paranti. Telugu: Bandhukamu, Manmadhabanam.
Trade name: Jungle Flame Ixora, Flame flower.
Propagation: by seeds as well as through stem cuttings.

Medicinal uses: Roots are used for treating fever, gonorrhoea, hiccup, anorexia, diarrhoea, dysentery, sores, chronic ulcers and skin diseases. Flowers are used to treat leucorrhoea, haemoptysis, catarhal bronchitis, opthalmopathy, sores and ulcers. Leaves are used to treat diarrhoea.

Jasminum grandiflorum L.

Family: Oleaceae

Large climbing shrub. Leaves imparipinnate, leaflets 9, opposite, rhomboid-oblong. Flowers fragrant, white, in terminal corymbose panicles.
Cultivated for ornamental flowers.
Fl. & Fr.: June–December.
Sanskrit: Jati, Malati. Telugu: Jaji, Malati.
Trade name: Spanish Jasmine.
Propagation: by stem cuttings.

Meicinal uses: Whole plant is used in facial paralysis, mental debility, sterility, leprosy, skin diseases, fixing loose teeth, ulcers, corns and pruritus.

Jasminum sambac (L.) Ait.

Syn: *Nyctanthes sambac* L.

Family: Oleaceae.

Erect or sometimes small climbing shrub. Leaves broadly ovate or elliptic, shining. Flowers white, fragrant, in 3-flowered terminal cymes. Fruit globose berry, violet when ripe.
Cultivated for ornamental flowers.
Fl. & Fr.: March–August.
Sanskrit: Mallika. Telugu: Malli.
Trade name: Arabian Jasmine, Tuscan Jasmine.
Propagation: Layering.

Medicinal uses: A lotion made of the flowers is used for washing the face and eyes. A decoction of the leaves is used for fevers. Leaves are applied as a polutice for skin complaints and ulcers. Roots are used with leaves in eye lotions.

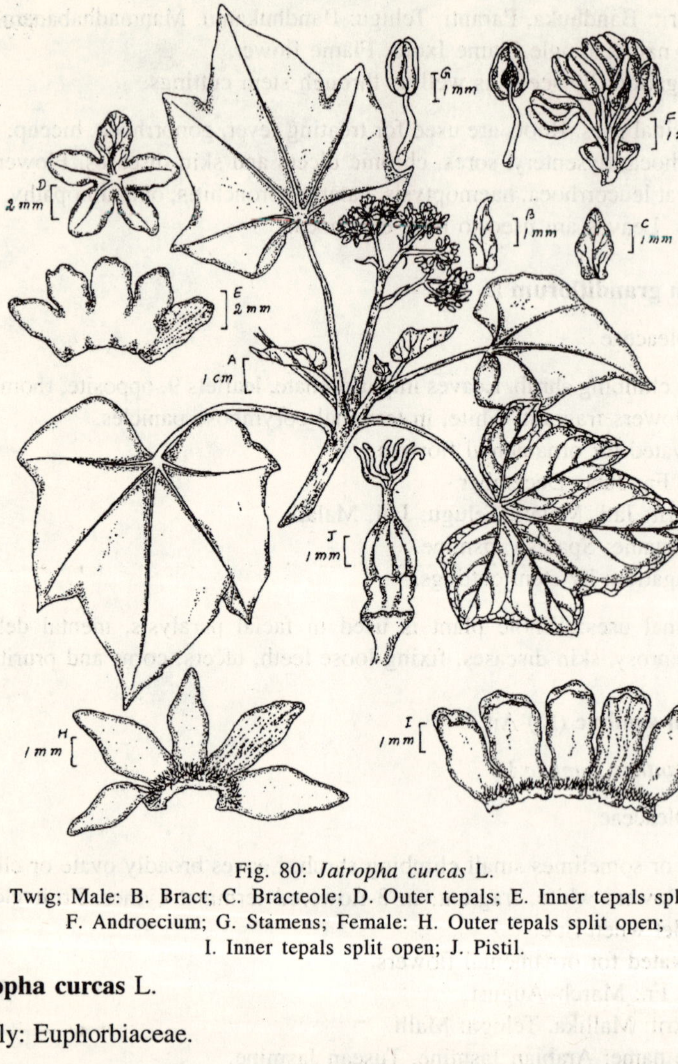

Fig. 80: *Jatropha curcas* L.
A. Twig; Male: B. Bract; C. Bracteole; D. Outer tepals; E. Inner tepals split open; F. Androecium; G. Stamens; Female: H. Outer tepals split open; I. Inner tepals split open; J. Pistil.

Jatropha curcas L.

Family: Euphorbiaceae.

Shrub or small tree, up to 5m tall. Leaves cordiform, shallowly 3–5-lobed. Flowers unisexual, monoecious, yellowish green in loose panicle of cymes. Fruit ovoid capsule.

Common hedge plant, along roadsides and outskirts of villages.

Fl. & Fr.: August–March.

Sankrit: Dravanthi, Hanaleronda. Telugu: Nepalamu, Adavi amudamu.

Trade name: Purging nut, physic nut.

Propagation: by cuttings and seeds.

Medicinal uses: The drug obtained from *J.curcas* is termed as "Dravanthi"

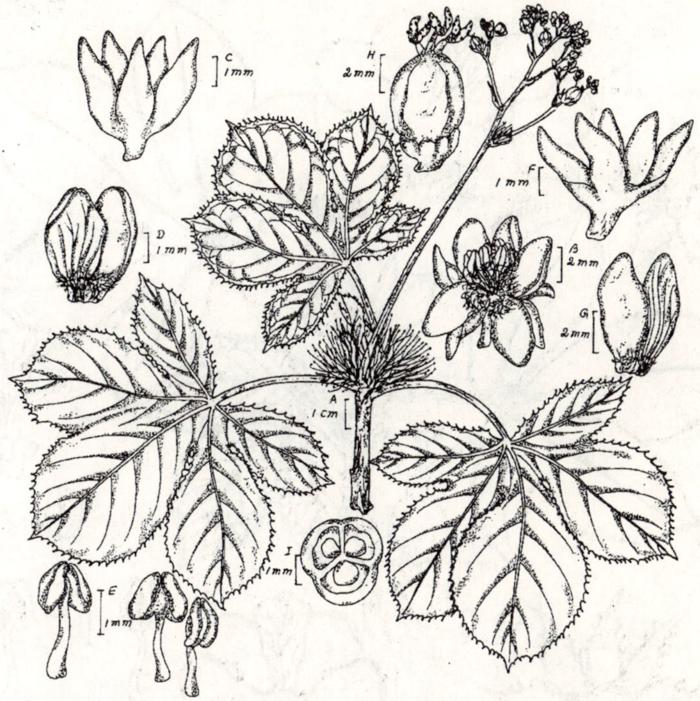

Fig. 81: *Jatropha glandulifera* Roxb.
A. Twig; Male: B. Flower; C. Outer tepals; D. Inner tepals (part); E. Stamens; Female: F. Outertepals; G. Inner tepals (part); H. Pistil; I. Ovary T.S.

and is reported to be bitter, pungent and astringent in taste. It has anthelmintic properties. It works as cleaner for all the impurities through its purgative properties. Besides this, it has been found beneficial in chronic dysentery, thirst, urinary discharges, abdominal complaints, anaemia, fistula, ulcer and disease of heart and skin. The leaves are considered to be rubifacient and lactogogue. Oil from the seed is used externally for rheumatism and paralytic affections.

Jatropha gladulifera Roxb.

Family: Euphorbiaceae.

Shrubs, to 2.5 m tall, branchlets stout, short. Leaves palmately 3–5-lobed, lobes obovate, serrate, serratures gland tipped. Flowers in axillarry cymes. Fruit capsule, suglobose.

A weed of waste places especially in black cotton soils.

Fl. & Fr.: August–January.

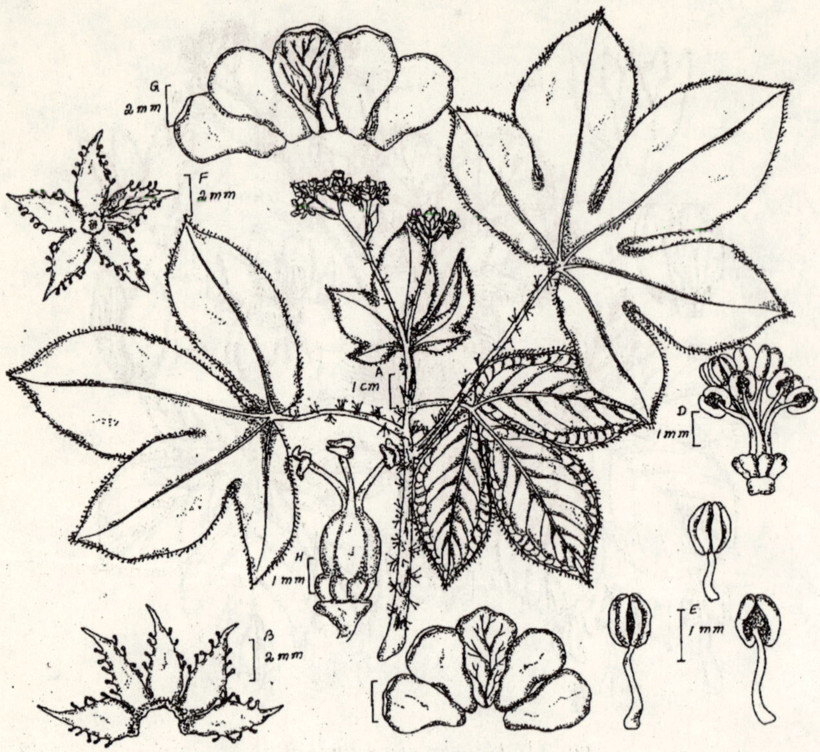

Fig. 82: *Jatropha gossypifolia* L. var. *elegans* (Pohl) Mucll.-Arg.
A. Twig; Male: B. Outer tepals split open; C. Inner tepals split open; D. Androecium; E. Stamens; Female: F. Outer tepals; G. Inner tepals split open; H. Pistil.

Sanskrit: Nikumbha. Telugu: Nepalamu, Adaviamudamu.

Propagation: by seeds.

Medicinal uses: The oil from seeds possess purgative properties. It is applied externally in rheumatism and paralytic affections.

Jatropha gossypifolia L.

Family: Euphorbiaceae.

Shrubs, to 3m tall, monoecious, branchlets dark purple, glandular hairy. Leaves glaucous, cordiform, deeply 3–5-lobed, lobes obovate. Flowers in axillary cymes. Fruit capsule, subglobose, seeds oblong.

Common in waste lands and roadsides.

Fl. & Fr.: Through out the year.

Sanskrit: Nikumba. Telugu: Seema Nepalamu, Nela Amudamu.

Trade name: Bellyache Bush.
Propagation: by seeds.

Medicinal uses. The dried stem bark of the plant contains an intensely bitter amorphous alkaloid, Jatrophine, which is similar to quinine in properties.

The roots are employed against leprosy. A decoction of the bark is an emmenagogue. The leaves are used to cure stomach ache, venereal diseases and as blood purifier. Seed oil is used to cure leprosy.

Lantana camara L. var. **aculeata** (L.) Moldenke

Syn: *Lantana aculeata* L.

Family: Verbenaceae.

Bushy scandent, prickly shrubs, up to 2.5 m tall. Leaves ovate, lamina rugose and scabrous. Flowers in umbellate spikes in various colours, yellow to dark crimson. Fruit drupe, shining black when ripe.

Common on road sides, along railway tracks and scrub jungles.
Fl. & Fr.: Through out the year.
Sanskrit: Caturangi, Vanacchedi. Telugu: Pulikampa.
Trade name: Wild sage. Lantana weed.

Medicinal uses: used in the treatment of itches and as antiseptic for wounds.

Active principle: The stem and root barks contain alkaloid lantanine with strong antipyretic and antispasmodic properties.

Lasia spinosa (L.) Thw.

Syn: *Dracontium spinosum* L.

A stout spinous marshy plant, rhizome thick, petiole, peduncle and leafy nerves beneath all prickly. Leaves when young hastate or sagittate, older ones often broader than long and deeply pedately pinnatifid, lobes linear-elliptic or oblong-lanceolate. Flowers in spadix enclosed in greenish brown spathe. Berries densely minutely muricate at apex, seeds rugose.

Rare in marshy localities in north coastal districts.
Fl. & Fr.: June–December.

Medicinal uses: The plant is recommended for colic, rheumatism and intestinal diseases. Root stock and fruits are used for affections of the throat. The juice of the root stock is given as a remedy for piles. Leaves are used for stomach ache and other pains.

Lawsonia inermis L.

Syn: *Lawsonia alba* Lam.

Family: Lythraceae.

Bushy shrubs, up to 3 m tall, branchlets spine-tipped, with greyish-white smooth bark.

Leaves opposite-decussate, obovate-elliptic. Flowers white or creamy-white in terminal panicles. Fruit globose capsule, seeds minute, brown.

Commonly cultivated all over the state for its leaves which yield dye.

Fl. & Fr.: February–June.

Sanskrit: Madyantika. Telugu: Gorintaku.

Trade name: Henna, Egyptian privet.

Medicinal uses: Leaves astringent and refrigerant. The drug is used externally for eczema, scabies, fungal infections and ulcers. It is also used for amoebic dysentery and gastrointestinal ulcers. It is also used as abortifacient. The drug is also contained in facial and hair lotions and is used to treat dandruf. Henna root preperations are used to treat leprosy, skin diseases, amenorrhoea and dysmenorrhoea. Henna leaves are used to treat wounds, ulcers, dysuria, coughs, bronchitis, one sided headache, rheumatism and anemia. The flowers are used for headache, fever and acute psychosis. Henna seeds are used to treat intermittent fever, diarrhoea and dysentery.

Lepidagathis cristata Willd.

Family: Acanthaceae.

Perennial procumbent herb, branches numerous from a highly reduced main stem. Leaves linear-lanceolate. Flowers white with brown or purple spots in globose heads, crowded at the base of the stem. Fruit oblong capsule.

Common in dry places in waste lands.

Fl. & Fr.: September–March.

Telugu: Mullabanti.

Trade name: Lepidagathis

Propagation: by seeds.

Medicinal uses: The plant is used as a bitter tonic in fevers. It is also applied to itchy affections of the skin.

Leptandenia reticulata (Retz.) Wight & Arn.

Family: Asclepiadaceae.

Twining shrub, with watery sap and deeply fissured bark. Leaves elliptic-oblong or lanceolate. Flowers greenish yellow in axillary umbellate cymes. Fruit thick woody follicles.

Common in hedges and bushes in waste lands and forests.

Fl. & Fr.: July–December.

Sanskrit: Jivanti, Arkapushpa. Telugu: Kalasa, Guttipala, Mukku tummudu.

Trade name: Cork swallow-wort, Jeevani.

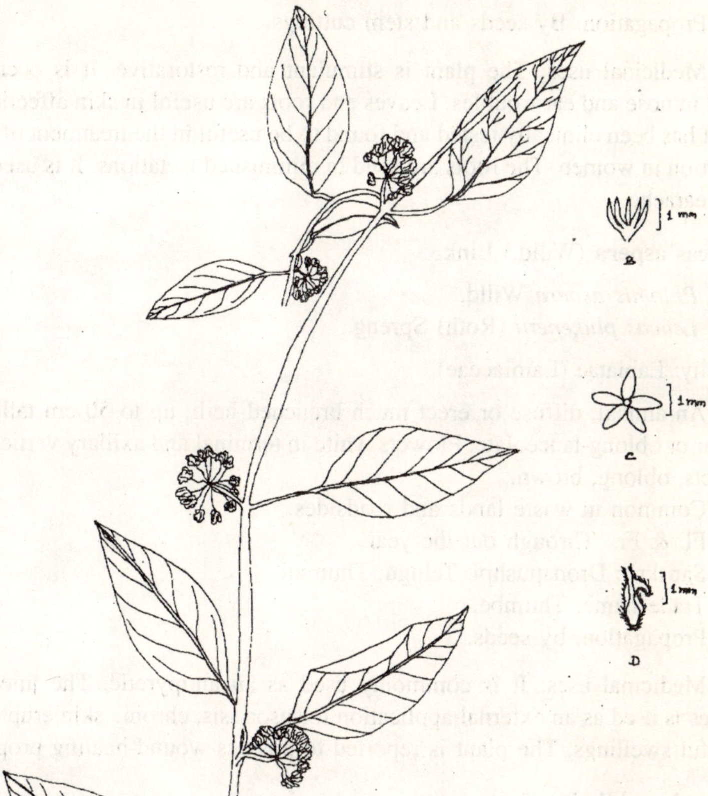

Fig. 83: *Leptadenia reticulata* (Retz.) Wight & Arn.
A. Twig; B. Calyx split open; C. Corolla split open; D. Corolla lobe with corona; E. Pollinia; F. Pistil; G. T.S of ovary.

Propagation: By seeds and stem cuttings.

Medicinal uses: The plant is stimulant and restorative. It is occasionally used in nose and ear troubles. Leaves and roots are useful in skin affections. The plant has been clinically tested and found to be useful in the treatment of habitual abortion in women. The roots are used in diminished lactations. It is used in cold and earache.

Leucas aspera (Willd.) Link

Syn: *Phlomis aspera* Willd.
Leucas plukenetii (Roth) Spreng.

Family: Labiatae (Lamiaceae)

An annual, diffuse or erect much branched herb, up to 50 cm tall. Leaves linear or oblong-lanceolate. Flowers white in terminal and axillary verticils. Fruit nutlets, oblong, brown.

Common in waste lands and roadsides.
Fl. & Fr.: Through out the year.
Sanskrit: Dronapushpi. Telugu: Thummi.
Trade name: Thumbe.
Propagation: by seeds.

Medicinal uses: It is commonly used as an antipyretic. The juice of the leaves is used as an external application for psoriasis, chronic skin eruptions and painful swellings. The plant is reported to possess wound-healing properties.

Limonia acidissima L.

Syn: *Feronia limonia* (L.) Swingle
Feronia elephantum Correa
Limonia elephantum (Correa) Panigr.
Schinus limonia L.

Family: Rutaceae

Small armed tree, spines straight, branchlets densely tomentose. Leaves alternate, imparipinnate, leaflets 3–9, oblong or obovate, coriaceous, petiole and rachis flattened, narrowly winged. Flowers creamish in axillary and terminal panicles. Fruit globose berry, woody, indehiscent, seeds many, embedded in pulp.

Common in dry deciduous forests.
Fl. & Fr.: March–December.
Sanskrit: Kapittha. Telugu: Velaga.
Trade name: Wood apple.

Medicinal uses: The edible fruits are considered tonic and antiscorbutic. Leaves aromatic, carminative, astringent and yield an essential oil.

Litsea glutinosa (Lour) C.B.Robinson

Syn: *Litsea chinensis* Lamk.
Litsea sebifera Pers.
Sebifera glutinosa Lour.

Family: Lauraceae.

Small or moderate sized evergreen tree. Leaves alternate, elliptic to obovate. Flowers small, white or yellowish in few-flowered compound umbels. Fruit globose, purple when ripe.

Common in forests of north coastal, Cuddapah and Nellore districts.

Fl. & Fr.: June–April.

Telugu: Narra Alagi.

Propagation: by seed or coppice of shoots.

Medicinal uses: The bark of *Litsea glutinosa* constitutes the common demulcent drug sold in indian bazars under the name maida lakri or Maida lakadi. The leaves are mucilaginous and considered emollient and antispasmodic. They are used in infusion or as poultice for bruises and wounds. Flower buds are credited with similar properties. The root is sweetish bitter, astringent and tonic.

Luffa acutangula (L.) Roxb.

Syn: *Cucumis acutangula* L.
Luffa amara Roxb.

Family: Cucurbitaceae.

Extensive tendrillar climber. Tendrils often trifid. Leaves orbicular, palmately 5–7-angled, scabrid. Flowers pale yellow, male and female in the same axil. Fruit long, prominently 10-angled, seeds black, compressed.

Commonly cultivated for its fruits.

Sanskrit: Kosataki, Jalini. Telugu: Beerakai.

Trade name: Ribbed gourd.

Propagation: by seeds.

Medicinal uses: The plant possesses laxative and purgative properties and is reported to be useful in skin diseases and asthma. It is used as diuretic and in splenic enlargements. The dried fruit is powdered and used as snuff in jaundice. Seeds are considered emetic, expectorant and demulcent.

Luffa cylindrica (L.) M. Roem.

Syn: *Momordica cylindrica* L.
Luffa aegyptiaca Mill.

Family: Cucurbitaceae.

Monoecious slender climbing herbs, branchlets glabrous. Leaves 5–7-lobed, lobes ovate, scabrid, tendrils 3-fid. Flowers bright yellow, unisexual, male in short racemes, female solitary. Fruit cylindric, oblong, green with dark green longitudinal stripes, seeds many.

Cultivated for its fruit.
Fl. & Fr.: Round the year.
Sanskrit: Mahakosataki. Telugu: Netibeerakaya.
Trade name: Sponge gourd, vegetable sponge.
Propagation: by seeds.

Medicinal uses: Mature seeds are bitter emetic and cathartic. The seeds are used to cure syphilis and in haematuria and dermatopathy.

Macaranga peltata (Roxb.) Muell.-Arg.

Syn: *Osyris peltata* Roxb.
Macaranga roxbughii Wt.
Family: Euphorbiaceae.

Evergreen tree, up to 10 m tall, bark grey. Leaves orbicular, peltate, red glandular below. Flowers greenish yellow, in axillary panicles. Fruit globose capsule, glandular.

Occasional in high hills and moist valleys.
Fl. & Fr.: March–June.
Telugu: Konda tamara, Konda Japhra.
Trade name: Roxburgh's Lotus Croton.
Propagation: by seeds.

Medicinal uses: Gum which exudes from the cut branches, base of petioles, young shoots and fruits are used in external application for venereal sores.

Madhuca indica Gmel.

Syn: *Bassia latifolia* Roxb.
Madhuca latifolia Macb.
Family: Sapotaceae.

A deciduous tree, up to 20 m tall, branchlets tomentose. Leaves ovate. Flowers creamish white in axillary dense fascicles. Fruit berry, tawny tomentose, 1–4-seeded.

Common in all forests.
Fl. & Fr.: March–September.
Sanskrit: Madhuka, Vanaprastha. Telugu: Ippa.
Trade name: South India Mahua tree, Butter tree.

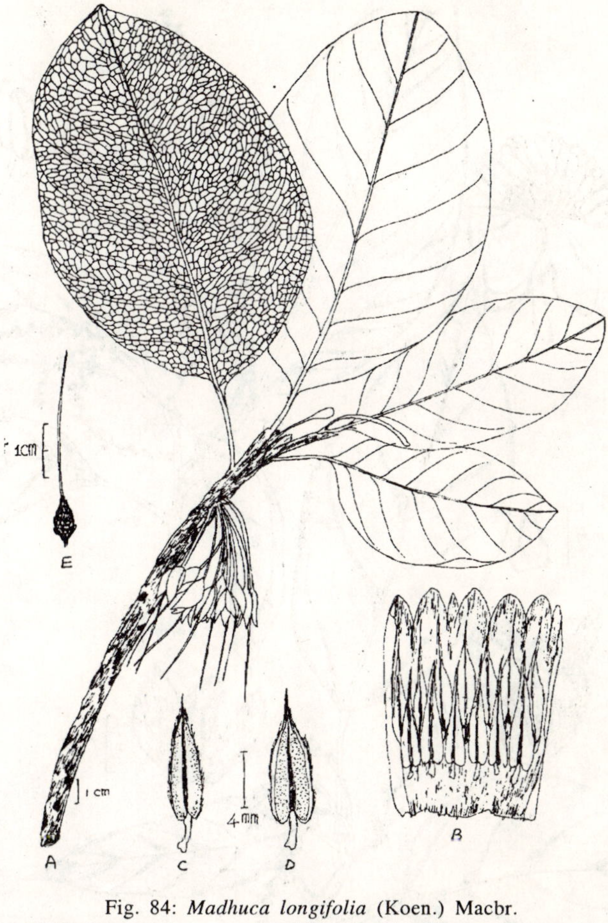

Fig. 84: *Madhuca longifolia* (Koen.) Macbr.
A. Twig; B.-E. Inner and outer calyx lobes; F. Corolla spread out with stamens; G. & H. Stamens; I. Pistil.

Medicinal uses: The bark, leaves, flowers and seeds of the tree constitute the drug. Decoction of the bark is applied in curing itches, bleeding gums and ulcers; it is given internally in diabetes. Leaves are astringent, their ash, mixed with butter is applied on burns and scalds. Flowers are used in cough and bronchitis; they are cooling and nutritive. A spirit prepared from flowers is considered to be tonic and nutritive. Flowers possess some antibacterial property. Seeds promote formation and flow of milk. Oil from seeds is good on skin diseases and laxative.

Madhuca longifolia (Koenig) Macbr.

Syn*: Bassia longifolia.*

Family: Sapotaceae.

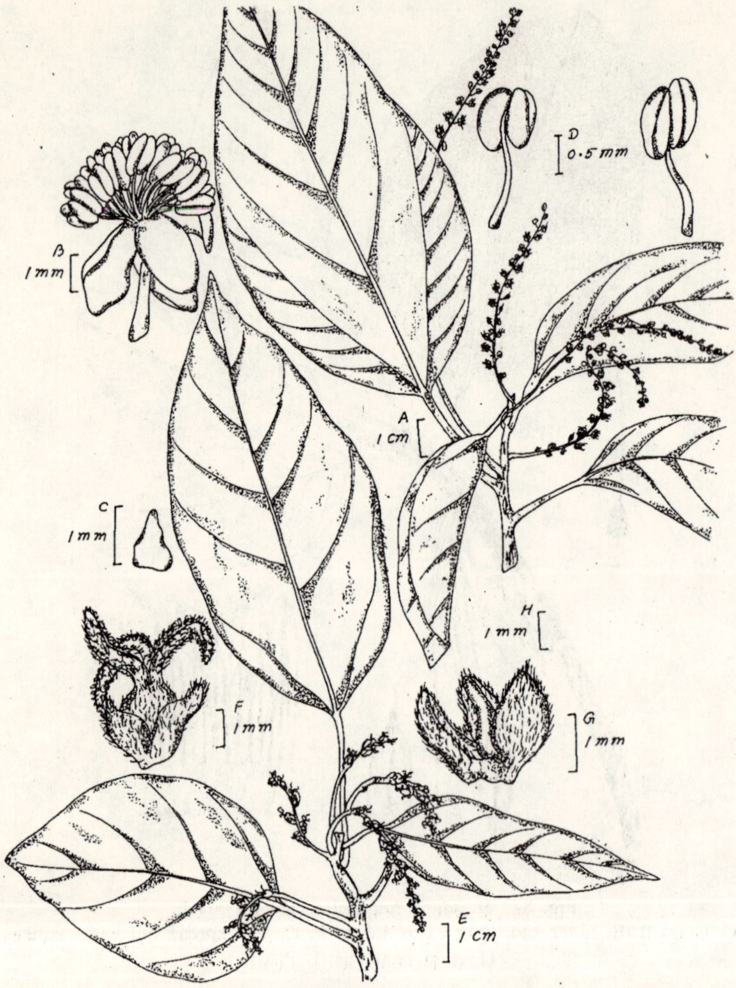

Fig. 85: *Mallotus philippensis* (Lam.) Muell.-Arg.
A. Twig; Male: B. Flower; C. Bract; D. Stamens; Female: E. Twig; F. Flower; G. Tepals; H. Pistil.

Deciduous trees, up to 20 m tall, bark dark brown. Leaves clustered near the ends of the branches, linear-lanceolate. Flowers creamish white in terminal umbels. Fruit berry, ellipsoid, 1-seeded.

Occasional in forests.

Fl.: March. Fr.: May–June.

Sanskrit: Madhula, Gudapushpa. Telugu: Ippa, Peddaippa.

Trade name: South Indian mahua, Butter tree.

Propagation: by seeds.

Medicinal uses: Bark is reported to be good for inflammations and sprains. Heartwood is recommended for epilepsy. Flowers are stimulant, diuretic and reportedly used for treating diseases of the stomach. Fruits are used as aphrodisiac and also for treating bronchitis. The oil from the seed is laxative and reported to be used in dermatopathy, rheumatism and haemorrhoids.

Mallotus philippensis (Lam.) Muell-Arg.

Family: Euphorbiaceae.

Small or moderate sized tree, up to 12 m tall, branchlets grey-tomentose. Leaves elliptic-ovate, tomentose with orbicular red glands below. Flowers unisexual, male flowers in axillary or terminal short spikes. Fruit capsule, red glandular, seeds subglobose, black.

Common in forests.

Fl. & Fr.: October–February.

Sanskrit: Kampillaka, Kamela, Kapila. Telugu: Sinduramu, Kunkuma Chettu.

Trade name: Kamala Dye tree.

Propagation: by seeds.

Medicinal uses: The red glandular and hairy substance on fruits is used for destroying tape worms. Glandular hairs of the fruits are also used in Eye diseases, leprous eruptions, ascites, diabetes, vermicosis, constipation, flatulence, wounds, ulcers, cough, scabies, ringworm, herpes and haemorrhages.

Mangifera indica L.

Family: Anacardiaceae

Large, spreading, evergreen trees. Leaves crowded at the ends of branches, coriaceous, oblong-lanceolate, glabrous and shining. Flowers whitish in many-flowered pubescent panicles. Fruit drupe, ovoid or obliquely pyriform, 1-seeded.

In moist forests, also cultivated for its edible fruit.

Fl.: January–March; Fr.: April–July.

Sanskrit: Amrah, Chutah. Telugu: Mamidi.

Trade name: Mango tree.

Propagation: by seeds and grafting.

Medicinal uses: The fruit is a rich source of Vitamin A and C apart from minerals and other Vitamins. The bark is astringent, it is used in diphtheria, and rheumatism. It possesses a tonic action on the mucous membrane. The kernal juice is snuffed once a day for three days to stop nasal bleeding. The unripe mango is useful in ophthalmia and eruptions. The rind is astringent, stimulating tonic in debility of stomach.

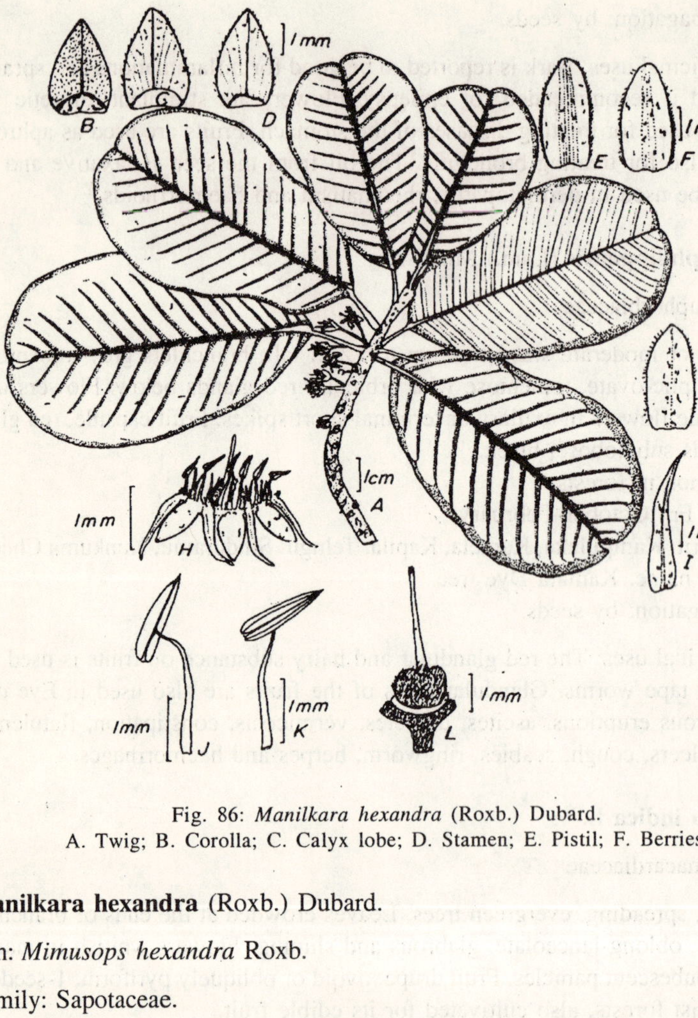

Fig. 86: *Manilkara hexandra* (Roxb.) Dubard.
A. Twig; B. Corolla; C. Calyx lobe; D. Stamen; E. Pistil; F. Berries.

Manilkara hexandra (Roxb.) Dubard.

Syn: *Mimusops hexandra* Roxb.

Family: Sapotaceae.

Evergreen tree, bark blackish grey, deeply furrowed. Leaves forming a dense crown, elliptic-obovate, glabrous. Flowers axillary, solitary or in fascicles. Fruit ellipsoid berry, seed solitary.

In Dry ever green forests.

Fl.: & Fr.: July–February.

Sanskrit: Rajadanah. Telugu: Pala.

Trade name: Obtuse leaved Mimusops.

Propagation: by seeds.

Medicinal uses: The stem bark is used as tonic and astringent.

Martynia annua L.

Syn: *Martynia diandra* Gloxin.

Family: Pedaliaceae.

Annual, glandular pubescent, erect undershrubs, up to 1 m tall. Leaves ovate-cordate, pubescent and sticky. Flowers showy, cream-colorued, in axillary racemes. Fruit woody with two sharp hooks at the apex, pericarp green, sticky, mesocarp black.

Common in waste lands, road sides and hedges of fields.

Fl. & Fr.: July–December.

Telugu: Telukondi kaya.

Trade name: Devil's claw, Tiger claw.

Propagation: by seeds.

Medicinal uses: The leaves are used in epilepsy and applied to tuberculous glands of the neck. Leaf juice is used as a gargle for sore throat. The fruit is considered alexiteric and useful in inflammations.

Chemistry: Leaves contain chlorogenic acid.

Melia azedarach L.

Family: Meliaceae.

Medium sized tree, up to 10m tall. Leaves 2–3-pinnate, leaflets 1–5 pairs, opposite, ovate-obovate or lanceolate, glabrous, base oblique. Flowers lilac in long panicles. Fruit ellipsoid drupe, fleshy, seed solitary.

Cultivated all over the state and occasionally found run wild.

Fl. & Fr.: April–December.

Sanskrit: Himadruma, Parvatanimba Vruksha. Telugu: Thurakavepa, Kondavepa.

Trade name: Persian lilac.

Propagation: by seeds.

Medicinal uses: The leaf is considered anthelmintic, diuretic, antilithic and emmenagogue. A decoction of the leaves is regarded as astringent and stomachic. A poultice of the flowers is applied to eruptive skin diseases and killing lice. A gum collected from the tree is considered useful in spleen enlargement. The plant is sedative and psychostimulant.

Memecylon edule Roxb.

Family: Melastomataceae.

Densely foliaceous, evergreen tree, bark deep brown. Leaves ovate, coriaceous, glabrous. Flowers blue, numerous in clustered, compound, umbellate cymes. Fruit globose berry, blue or dark purple when ripe.

Occasional in ravines in forests.
Fl.: & Fr.: April–July.
Sanskrit: Anjani. Telugu: Alli.
Trade name: Iron wood Tree.
Propagation: by seeds.

Medicinal uses: Roots are used in menorrhagia and heavy menstruation. Leaf decoction is used in washing the eyes.

Mentha spicata L.

Family: Lamiaceae (Labiatae).

Perennial, stoloniferous herb, 30–60 cm tall. Leaves oblong-ovate or lanceolate, decussate. Flowers pale lilac, pink or white in false whorls in spike like inflorescesnce. Fruit nutlet.
Cultivated for its leaves.
Telugu: Pudina.
Trade name: Spearmint.
Propagation: by stolons.

Medicinal uses: It is used as carminative. The dried tops are used as a stimulant and to prevent vomitting. Tea made from this plant is used to treat neuralgia, indigestion and diarrhoea and also to wash sores. Tea made from leaves with cinnamon is prescribed to expectant mother to hasten child birth.

Merremia gangetica (L.) Cuf.

Syn: *Merrenia emarginata* (Burm.f.) Hallier f.
Convolvulus gangeticus L.
Evolvulus emarginatus Burm.f.
Ipomoea reniformis (Roxb.) Choisy

Family: Convolvulaceae.

A perennial, slender, diffuse, creeping herb, rooting at nodes. Leaves reniform, alternate. Flowers yellow, solitary or 2–3-flowered cymes, axillary. Fruit globose capsule surrounded by slightly enlargedd sepals, 2-seeded.
Common weed of moist waste lands and marshy places.
Fl. & Fr.: October–January.
Sanskrit: Mushakarni. Telugu: Elika Jeevaku, Elika jemudu.
Propagation: by seeds.

Medicinal uses: Whole plant is used in Nephropathy, uropathy, pneumonosis, cardiac diseases, metropathy, strangury, otalgia, cephalagia and rat bite. The plant is considered deobstruent, diuretic and alterative and used in rheumatism and neuralgia.

Merremia tridentata (L.) Hallier f.

Syn: *Convolvulus tridentatus* L.
Ipomoea tridentata (L.) Roth

Family: Convolvulaceae.

A perennial spreading herb with thick root stock. Leaves linear-ovate, hastate. Flowers creamy-yellow with maroon centre, solitary or in few-flowered cymes. Fruit capsule, depressed–globose, 4-seeded.

Common in waste lands and roadsides.

Fl. & Fr.: June–March.

Sanskrit: Prasarani. Telugu: Sitasavaram, Seethamma jada.

Propagation: by seeds.

Medicinal uses: The plant is considered bitter, astringent calefacient and tonic and used in the treatment of rheumatism, hemiplegia, piles, swelling and urinary disorders. A decoction of the root is also used for the same purpose.

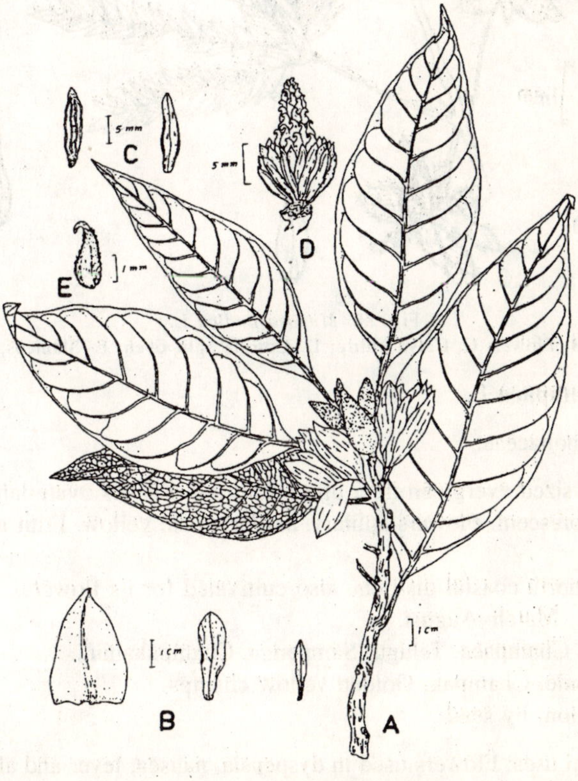

Fig. 87: ***Michelia champaca*** L.
A. Twig; B. Perianth lobes; C. Stamens; D. Stamens with pistil; E. Pistil.

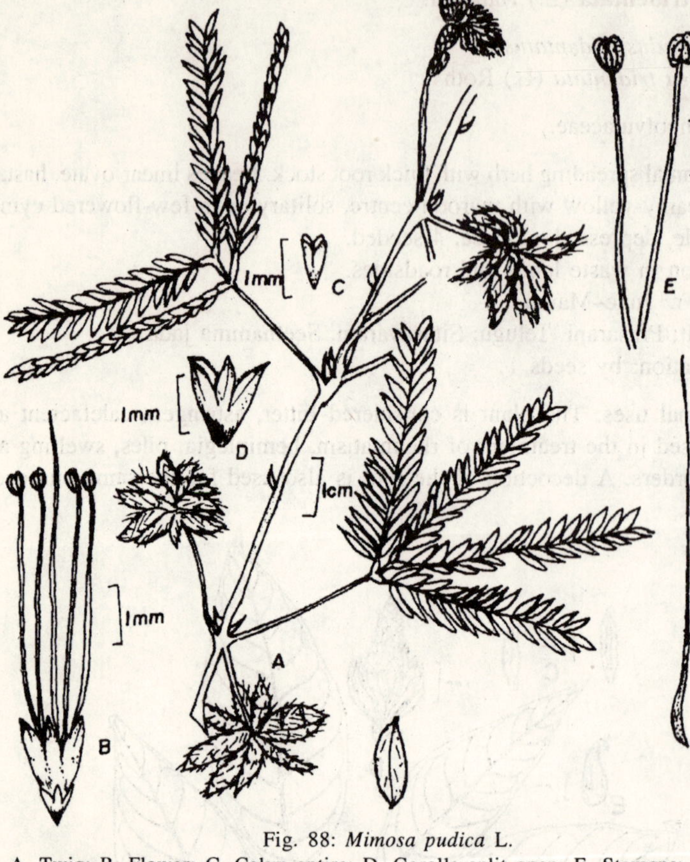

Fig. 88: *Mimosa pudica* L.
A. Twig; B. Flower; C. Calyx entire; D. Corolla split open; E. Stamens; F. pistil.

Michelia champaca L.

Family: Magnoliaceae.

Medium sized evergreen tree, up to 20m tall. Leaves ovate-lanceolate, coriaceous, glabrescent. Flowers solitary, axillary, pale yellow. Fruit an aggregate of follicles.

Rare in north coastal districts, also cultivated for its flowers.

Fl. & Fr.: March–August.

Sanskrit: Champaca. Telugu: Sampenga, Champakamu.

Trade name: Champak, Golden yellow champa.

Propagation: by seeds.

Medicinal uses: Flowers used in dyspepsia, nausea, fever and also useful as a diuretic and febrifuge. The bark is diuretic, febrifuge and stimulant. The oil extracted from flowers is used in cephalagia, opthalmia, gout and rheumatism.

Root juice is abortifacient. Dry roots and root bark are used as purgative and emmenagogue.

Mimosa pudica L.

Family: Mimosaceae.

Spreading herbs, branchlets glabrescent or hispid, prickles short, erect or curved. Leaves up to 4 cm long, pinnae 1–2 pairs, leaflets 15–20 pairs, elliptic-oblong, overlapping. Leaves sensitive to touch. Flowers pink in axillary heads. Fruit flat pod, slightly undulate, 2–4-jointed, bristly along margins, seeds ovoid.

Common in waste lands and road sides, especially in moist localities.

Fl. & Fr.: July–November.

Sanskrit: Lajjalu, Namaskari. Telugu: Attipatti, Lajja nathi.

Trade name: Sensitive plant.

Propagtion: By seeds.

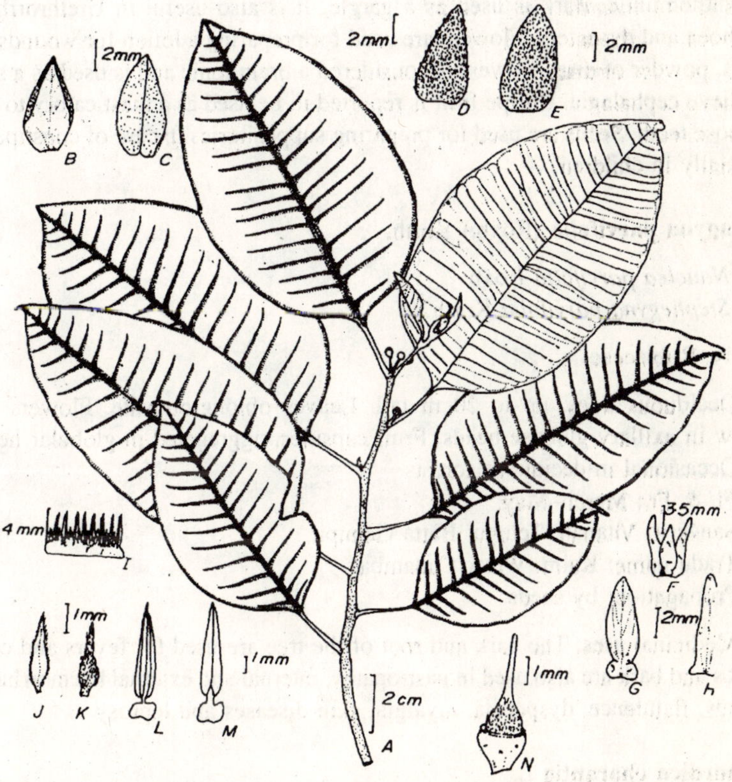

Fig. 89: *Mimusopos elengi* L.
A. Twig; B.-E. Sepals; F. Part of Corolla; G. & H. Corolla lobes; I. Androecium; J. & K. Staminodes; L. & M. Stamens; N. Pistil

Medicinal uses: A decoction of the root is taken to relieve asthma and diarrhoea. The plant is used in menorrhagia, piles, minor skin wounds, diarrhoea, fistula, hydrocele, piles, diabetes filariasis and whooping cough.

Chemistry: *Mimosa pudica* contains toxic alkaloid mimosine.

Mimusops elengi L.

Family: Sapotaceae.

Medium sized evergreen trees, up to 10 m tall, bark scaly. Leaves elliptic-oblong. Flowers solitary or few in fascicles of 2–6. Fruit ovoid berry.

Deccan in ravines in the hills, also cultivated.

Fl. & Fr.: March–May.

Sanskrit: Bakula, Sinhakesara. Telugu: Pogada.

Propagation: by seeds.

Medicinal uses: The bark, flowers and fruits are acrid, astringent, cooling and anthelmintic. Bark is used as a gargle. It is also useful in Urethrorrhoea, diarrhoea and dysentery. Flowers are used for preparing a lotion for wounds and ulcers, powder of dried flowers is considered a brain tonic and is used as a snuff to relieve cephalagia. Unripe fruit is reported to be used as a masticatory to help fix loose teeth. Seeds are used for preparing suppositories in case of constipation especially in children.

Mitragyna parvifolia (Roxb.) Korth.

Syn: *Nauclea parvifolia* Roxb.
Stephegyna parvifolia Korth.

Family: Rubiaceae.

Deciduous trees, up to 20 m tall. Leaves oblong-obovate. Flowers pale yellow in axillary globose heads. Fruit capsules, aggregated in globular head.

Occasional in deciduous forests.

Fl. & Fr.: March–May.

Sanskrit: Vitanah. Telugu: Batta Genupu.

Trade name: Kaim, Water Cadamba.

Propagation: by seeds.

Medicinal uses: The bark and root of the tree are used for fevers and colic. Leaves and bark are also used in gastropathy, internal and external haemorrhages, wounds, flatulence, dyspepsia, myalgia, skin diseases and leprosy.

Momordica charantia L.

Family: Cucurbitaceae.

Annual, pubescent, tendrillar climber. Leaves alternate, simple cordate, much lobed. Flowers in leaf axils, monoecious. Fruit spindle-shaped with wart like surface, seeds with red aril.

Cultivated for fruit vegetable.

Sanskrit: Karvellam, kandakataka. Telugu: Kakara.

Trade name: Bitter gourd.

Propagation: by seeds.

Medicinal uses: The fruits are considered tonic, stomachic, carminative and cooling, they are used in rheumatism, gout and diseases of liver and spleen. The fruits, leaves and roots have long been used in India as a folk remedy for diabetes mellitus. It is claimed that powdered fruit is effective in healing wounds.

Monochoria hastata (L.) Solms-Laub.

Syn: *Pontederia hastata* L.

Monochoria hastaefolia Presl.

Family: Pontederiaceae.

A fresh water, floating erect herb, root creeping, rooting below, spongy. Leaves sagittate, hastate or cordate. Flowers brilliant purplish blue, crowded, racemose or subumbellate. Fruit capsule, oblong, many-seeded.

In ponds and tanks, rare.

Fl. & Fr.: August–December.

Sanskrit: Nilotpalam.

Propagation: by seeds.

Medicinal uses: The plant is considered alterative, tonic and cooling. A juice of the leaves is applied to boils and also used in insanity. Root stock increases vitality. It is used as a remedy in sensation in the heart and blood.

Monochoria vaginalis (Burm.f.) Presl.

Syn: *Pontederia vaginalis* Burm.f.

Family: Pontederiaceae

A rooted aquatic herb, root stock spongy. Leaves variable, linear to ovate or ovate-cordate. Flowers blue, spotted with red, in axillary raceme. Fruit capsule, oblong.

Occasional in ponds, tanks and streams.

Fl. & Fr.: August–December.

Sanskrit: Indivarah. Telugu: Nirkacha.

Propagation: by seeds.

Medicinal uses: Root stock and leaves are medicinal. They are used in burning sensation of the body, dipsia, stranguary, gastropathy, hepatopathy, asthma, scurvy, boils and haemorrhages.

Morinda citrifolia L.

Family: Rubiaceae.

Shrubs, up to 3 m tall. Leaves coriaceous. Flowers white, in dense ovoid heads. Fruit syncarpium, white when ripe.

Occasional in hills of coastal districts.

Fl. & Fr.: July–October.

Sanskrit: Asyaka. Telugu: Togara.

Trade name: Indian mulberry.

Propagation: by seeds.

Medicinal uses: The root is used as a cathartic and febrifuge, and applied externally to relieve pain in gout. Leaves are considered tonic and febrifuge. They are used as healing application for wounds and ulcers. The juice of the leaves is externally applied in gout. Fruits are used for spongy gums, throat complaints, dysentery, leucorrhoea and sapraemia. A decoction of the root is taken to regulate menstruation. The root extract is given to relieve hyper tension.

Moringa concanensis Nimmo ex Gibs.

Family: Moringaceae.

Deciduous tree, up to 10 m tall, bark deeply fissured. Leaves bipinnate, pinnae 5–8 pairs, leaflets 4–6 pairs, orbicular or elliptic. Flowers white with purple streaks in terminal panicles. Fruit elongated capsule, seeds many with papery membranous wings.

Occasional in forests.

Fl. & Fr.: February–August.

Sanskrit: Sigru, Bahulada. Telugu: Konda munaga, Adavimunaga.

Trade name: Soanjana

Propagation: by seeds.

Medicinal uses: Similar to *Moringa pterigosperma*.

Moringa pterigosperma Gaertn.

Syn: *Moringa oleifera* auct non Lam.

Family: Moringaceae.

Small trees. Leaves tripinnate, leaflets small, ovate, elliptic-obovate. Flowers white, strongly honey scented, in panicles. Fruit elongated capsule, loculicidally dehiscent, seeds rounded, winged.

Cultivated through out the state for its fruit.

Sanskrit: Sigruh, Sobhanjanah, Madhugunjana. Telugu: Munaga.

Trade name: Drumstick tree.

Propagation: by seeds, cuttings

Medicinal uses: The root has been used internally in folk medicine for gastrointestinal complaints, epilepsy, paralyses, cardiac and blood pressure disturbances, fever, scurvy, dizziness and colds. External indications include gingivites, worm diseases, snake bites, absceses, inflamation, rheumatism and pooly, healing wounds. Root paste has been used to treat worms, rheumatism and headaches. The seeds are used in neuralgia, inflamations, intermittent fevers and opthalmopathy. Flowers are used as aphrodisiac. The leaves are rich in vitamin A and C and are used as anti-inflamatory, anodyne, anthelmintic and opthalmic.

Active principle: Glucosinolates.

Mucuna pruriens (L.) DC.

Syn: *Dolichos pruriens* L.
Mucuna prurita Hook.
Carpopogon pruriens L.

Family: Leguminosae (Fabaceae).

Twining herbs, branches terete, striate, downy pubescent. Leaves trifoliolate, leaflets ovate-rhomboid, appressed white-pubescent. Flowers dark purple with yellow shade on standard petal in long axillary racemes. Pods curved, prominently S-shaped, densely silky pubescent with persistent pale brown or grey irritant bristles, seeds 4–6, orbicular.

Occasional in damp places, bushes, hedges and ravines in Telangana and north coastal Andhra Pradesh.

Fl. & Fr.: October–April.

Sanskrit: Atmagupta, Vanari. Telugu: Dulagondi.

Trade name: Common cowitch.

Propagation: It can be propagated by seeds. *Mucuna pruriens* is being cultivated all over India. The crop can be grown on a wide range of soils but it does well on sandy to clay loam type of soils, provided drainage is adequate. This species can be grown as a rainfed crop as well as irrigated crop. As a rainfed crop, sowing is done in June and final harvesting is done in December. As irrigated crop it is not season bound. The best season for planting is August and harvesting is done in January and February. A basal dose of 13–15 tonnes of Farm yard manure along with 80 kg P_2O_5 per hectare should be applied to the fields before ploughing. Top dressing of a mixture of 100 kg of Nitrogen and 75 kg of Potash is given in two equal split doses when the crop is 30 days old and 60 days old.

The fields are ploughed with a mould broad plough 2–3 times till a fine belt is obtained. Basal dose of manure along with phosphate should be mixed in the soil at this time. Ridges are made along the contours with a spacing of 60 cm. The seed rate is 50 kg per hectare. 60 x 60 cm is provided if support is given

Fig. 90: *Mucuna pruriens* (L.) DC.
A. Twig; B. Calyx split open; C. Standard petal; D. Wing petal; E. Keel petal; F. Staminal column; G. Pistil.

to the plants, 60 × 40 cm if no support is given to the plants. The crop requires 1–2 weedings in the initial stages.

Leaves may be attacked by the Cercospora leaf spot. One spray of Dithane at 1.5 ml/litre at 30 days of growth are adequate as a prophylactic measure. The flowering in this crop is continuous and the pods mature periodically. The flowering starts after about 45 days growth. The dried pods should be harvested three or four times in the season. The pods are sun dried. The pods will break open on drying. These are to be winnowed and seperated from the husk.

The crop yield is about 1500–1750 kg /ha if no support is given to plants. If support is provided the yield will be 3000–3750 kg/ha.

Medicinal uses: The seeds of *Mucuna pruriens* contain about 5–6% of L-Dopa, 12.5% lecithins and alkaloids. L-Dopa, an aminoacid (L-hydroxy phenylalanine) is highly reputed medicine for curing Parkinson's diseases. L-Dopa is also used as fracture healing promoter and anti depressant drug. It's natural source is shown to give better protection and long term benefits as compared to synthetic Dopa. Recently Zandu Pharma Ltd. Bombay has marketed a standardized processed seed powder of *Mucuna pruriens* under the trade name H P-200 for all sorts of Parkinsonia.

The ancient Ayurvedic system of medicine has described the medicinal use of the seed for the treatment of diseases of central nervous system and also as a geriatric tonic.

The bristly hairs covering fresh and dry pods cause intense itching on contact with skin and sometimes cause blisters and dermatitis. A wash with warm water containing alkali or ammonia reduces the burning sensation. The hairs were formerly used as vermifuge and were official in some pharmacopoeias. They are available in bazaars as yellowish brown loosely felted masses, occasionally containing portions of the black pericarp and administered as a bolus (with bat) or electuary (with syrup, honey or treacle), followed by purgative. The itching caused by the trichomes of the pod is attributed to the presence of histamine-liberating principle.

Murraya koenigii (L.) Spreng.

Syn: *Bergera koenigii* L.

Family: Rutaceae.

Shrubs or small trees. Leaves pinnately compound, leaflets alternate, asymmetrical, oblong-lanceolate. Flowers white, small, in terminal corymbose panicles, fragrant. Fruit subglobose berry, purplish black when ripe, seeds 1 or 2.

Common on hill slopes in forests. Cultivated for its aromatic leaves, which are used for curry flavouring.

Fl. & Fr.: March–August.

Sanskrit: Kalasaka, Surabinimbha. Telugu: Karivepaku.

Trade name: Curry leaf tree.

Medicinal uses: Roots, bark and leaves are useful in hyperdipsia, burning sensation, pruritus, leprosy, skin diseases, anorexia, helminthiasis, dyspepsia, colic, flatulence, diarrhoea, dysentry, vomitings and stomach ache.

Chemisty: The leaves contain high amount of oxalic acid. The leaves also contain crystalline glucoside, koenigin and a resin.

Musa paradisiaca L.

Family: Musaceae.

Perennial, rhizomatous, tall erect herb, its robust tree like stem is composed of the sheathing spiral leaf bases. At the summit of 3–8 m stem a crown of large oval deep green leaves are produced, leaves up to 3 m long. Flowers in a fleshy spathe like scales which drop off as the fruit matures. Fruit fleshy, indehiscent.

Cultivated for its fruit.

Sanskrit: Kadali, Vanalakshmi. Telugu: Aratikaya.

Trade name: Banana, Plantain.

Medicinal uses: The juice of the flower mixed with curds is used in dysentery and excessive bleeding during menstruation in young girls. The juice mixed with a glassful of cucumber juice is given as a medicine to cure scanty urination and gonorrhoea. The sap of the stem with tender coconut water is given as a medicine for hysteria, epilepsy nervous insomina etc. A glassful of the plain sap is drunk in dysentery, diarrhoea, piles etc. Banana flower is eaten as a vegetable and medicine for diabetes, chronic cough, dyspepsia, constipation etc. Banana fruit is used as tonic, gastrointestinal disorders, in diabetes, chronic cough, dyspepsia, constipation etc.

Naravelia zeylanica (L.) DC.

Family: Ranunculaceae.

Climbing herbs. Leaves 3-foliolate, terminal leaflet modified as a tendril, lateral leaflets ovate. Flowers greenish-yellow in axillary and terminal panicles. Fruit achene, prolonged into a feathery style which persists in fruits as twisted pilose tail.

Common in coastal and Rayalaseema forests.

Fl.& Fr.: Septermber–March.

Sanskrit: Dhanavalli. Telugu: Pulla Bachala.

Propagation: by seeds.

Medicinal uses: Whole plant is used in rheumatism, helminthiasis, dermathopathy, odontalgia, cephalalgia, colic inflammations, wounds and ulcers and leprosy.

Nelumbo nucifera Gaertn.

Syn: *Nelumbium speciosum* Willd.
Nymphaea nelumbo L.

Family: Nelumbonaceae

Large perennial aquatic herb, with stout creeping underground rhizome. Leaves peltate, rotund, radiately nerved, waxy gaucous beneath. Flowers rose red, white or yellow.

Common in tanks and ponds all over the state.

Fl. & Fr.: Through out the year.

Sanskrit: Pankaj. Telugu: Kamalam, Tamara.

Trade name: Sacred Lotus, East Indian Lotus.

Medicinal uses: Flower is used for cholera, diarrhoea, worm infestation, vomiting, states of exhaustion and intermittant fevers. The rhizomes are applied as a paste in ring worm and other cutaneous affections. The plant is also used for intrauterine growth retardation, menorrhagia, diarrhoea, dysentery and skin darkening.

Chemistry: Contains Isoquinoline alkaloids, Flavonoids and Tannins.

Nerium indicum Mill.

Syn: *Nerium oleander* Soland.
Nerium odorum L.

Family: Apocynaceae.

Large profusely branched evergreen, xerophytic shrubs, up to 3 m tall. Leaves linear-lanceolate, thickly coriaceous. Flowers rose red or white, sweet scented, in terminal, many-flowered polychasial cymes. Fruit paired follicles, cylindric, ribbed.

Cultivated through out Andhra Pradesh for its ornamental flowers.

Fl. & Fr.: Round the year.

Sanskrit: Karavirah, Asvamaraka. Telugu: Ganneru.

Trade name: Indian oleander.

Propagation: Vegetative propagation by means of stem cuttings.

Medicinal uses: Cardiac tonic, diuretic. The plant has been used as antidote, antibacterial, antileprotic, anticancer, cardiotonic and CNS depressent. The juice of tender leaves is reported to be good for opthalmia. Root causes abortion. The plant is highly poisonous.

Active principle: Various parts of the plant are known to yield cardenolides, pregnanes and triterpenes. The principal cardio-substance present in leaves is oleandrin.

Nicotiana tabacum L.

Family: Solanaceae.

Annual erect subshrubs. Leaves cauline, broadly ovate elliptic-ovate, upper leaves narrowed, lanceolate. Flowers in terminal corymboid panicles. Fruit capsule.

Cultivated for the tobacco leaf.

Sanskrit: Tamakhu. Telugu: Pogaku.

Trade name: Tobacco.

Propagation: by seeds.

Medicinal uses: Leaves are used in itches, sores, snake bites, cough, asthma and rheumatism.

Chemistry: Alkaloid nicotene.

Nyctanthus arbor-tristis L.

Family: Nyctanthaceae.

Bushy shrub or small tree to 4 m tall, rough all over with stiff white hairs; young branches sharply quadrangular, hairy. Leaves ovate to obovate, rough and scabrous above. Flowers white or yellow in capitate heads. Fruit compressed capsule.

Common in deciduous forests of Telangana and coastal Andhra pradesh.
Sanskrit: Parijatah, Sephalika. Telugu: Parijathamu.
Trade name: Night Jasmine, coral Jasmine.
Propagation: seeds.

Medicinal uses: Flowers for Sciatica, leaves for fever and rheumatism. The leaves are used in traditional medicine for the treatment of chronic fever, rheumatism and intestinal worm infection.

Active principle: Crystalline nyctanthin in flowers and alkaloids, resins and amorphous glycosides in leaves.

Nymphaea nouchali Burm.f.

Syn: *Nymphaea stellata* Willd.

Family: Nymphaeaceae.

Aquatic, rhizomatous, perennial herbs. Leaves long petioled, peltate, orbicular, pale purple below, green above. Flowers solitary, blue, white, rose or purple. Fruit berry, spongy, globose or ovoid, seeds numerous.
Common in ditches, ponds and tanks.
Sanskrit: Kumuda, Utpala. Telugu: Kaluva, Neetithamara, Allikada.
Trade name: Indian water lily.
Propagation: Rhizomes.

Medicinal uses: Rhizome is useful in diarrhoea, dysentery and general debility. The pulp of rhizome when applied externally acts as a rubefacient. Flowers are used as a cardiotonic. Seeds are used to treat diarrhoea.

Ocimum americanum L.

Syn: *Ocimum canum* Sims.

Family: Labiatae (Lamiaceae).

Erect much branched strongly aromatic herbs, stems grooved, pubescent. Leaves elliptic-ovate. Flowers in whorls, white or creamish in terminal racemes. Fruit nutlets, oblong, black.
Common in waste lands all over the state.
Sanskrit: Vanabarbarika, Aranyathulasi. Telugu: Kukkathulasi, Kuppathulasi.
Trade name: Hoary basil.
Propagation: by seeds.

Medicinal uses: The plant possesses aromatic, carminative, diaphoretic and stimulant properties. A decoction of the plant is taken for coughs, the leaves for dysentery. It is also used as mouth wash for relieving tooth ache. The juice of leaves is given to children for cold, catarrh and bronchitis. A paste of leaves is used as an external application for parasitical skin affections.

Ocimum basilicum L.

Family: Lamiaceae.

Erect herbs, up to 60 cm tall. Leaves elliptic-ovate or lanceolate. Flowers in whorls, white or slightly pale purple in simple or much branched thyrsoid racemes. Fruit nutlets, mucilaginous when wetted.

Common in waste lands.

Sanskrit: Barbari, Munjarik. Telugu: Bhutulasi, Vibudi patri.

Trade name: Sweet Basil, common basil.

Fl. & Fr.: Through out the year.

Propagation: by seeds.

Medicinal uses: The plant is useful in fever, cough, worms, stomach complaints and gout, juice of leaves is used as a nasal douche and is applied in skin diseases. Seeds given to cure dysentery and chronic diarrhoea. Leaves are used to cure earache, rheumatoid arthritis, anorexia, itching, amenorhoea, dysmenorrhoea and malaria.

Active principle: Essential oil, caffeic acid, Flavonoids.

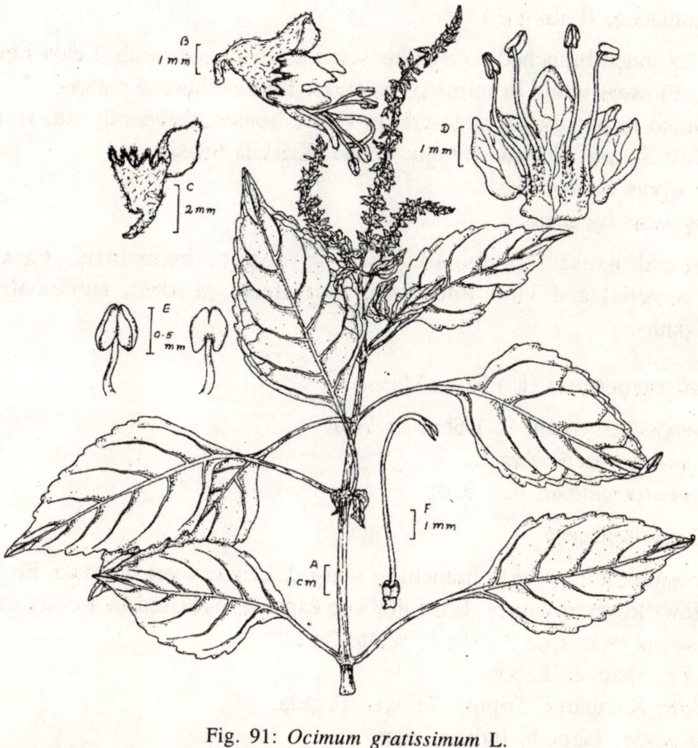

Fig. 91: *Ocimum gratissimum* L.
A. Twig; B. Flower; C. Calyx; D. Corolla split open; E. Anthers; F. Pistil.

Ocimum gratissimum L.

Family: Lamiaceae (Labiatae).

A much branched aromatic subshrub, up to 2 m tall. Leaves ovate, membranous. Flowers pale greenish yellow in close whorls in racemes. Fruit of 4 nutlets, subglobose, brown.

Occasional in waste lands and roadsides near villages.

Fl. & Fr.: October–March.

Sanskrit: Vriddhuthulasi. Telugu: Nimmathulasi, Lavanga thulasi.

Trade name: Shrubby Basil, Ban tulasi.

Propagation: by seeds.

Medicinal uses: Whole plant is used in Ear-ache toothache and abodominal colic in children. It is used as digestive tonic, stimulant, styptic, anthelmintic, neuralgia, urinary disorders, seminal weakness and paralysis.

Chemistry: Essential oil containing thymol and eugenol distilled from leaves.

Ocimum tenuiflorum L.

Syn: *Ocimum sanctum* L.

Family: Lamiaceae (Labiatae.)

An erect much branched softly pubescent aromatic subshrub. Leaves elliptic-oblong. Flowers white in terminal racemes. Fruits ellipsoid nutlets.

Cultivated in pots or on pedestals in Hindu houses, frequently run wild.

Sanskrit: Surasa, Vrinda. Telugu: Tulasi, Krishna tulasi.

Trade name: Sacred Basil.

Propagation: by seeds.

Medicinal uses: Leaves are used in cough, bronchitis, catarrh, halitosis, bacterial and viral infections, foul ulcers, wounds, anorexial and ophthalmopathy.

Operculina turpethum (L.) Silva Manso.

Syn: *Merremia turpethum* (L.) Shah & Bhat
Convolvulus turpethum L.
Ipomoea turpethum (L.) R.Br.

Family: Convolvulaceae

A perennial, stout twiner, branchlets winged. Leaves ovate-cordate. Flowers white in few-flowered cymes. Fruit globose capsule, enclosed in woody calyx.

Occasional twiner on shrubs in waste lands.

Fl.& Fr.: October–February.

Sanskrit: Kalaparni, Triputi. Telugu: Tegada.

Trade name: Turpeth, Indian Jalap.

Propagation: by seeds.

Medicinal uses: The dried roots of the plant constitute the drug, the roots should be taken from white variety only, and they should have thick bark intact. The Turpeth has almost same properties as the true Jalap, obtained from *Exogonium purga* and a suitable substitute for it. It is used as a purgative; powdered roots are suitable for this.

Active principle: Turpethin from roots.

Opuntia stricta (Haw.) Haw.

Syn: *Cactus strictus* Haw.
Opuntia dillenii Ker-Gawl.

Family: Cactaceae

Erect, branching, leafless xerophtic shrubs, stem jointed, joints flat, obovate fleshy with many areoles bearing yellowish-spines. Flowers yellow, solitary, near the margins of joints. Fruit berry, globose, fleshy.

Common in waste lands and road sides.

Fl. & Fr.: January–August.

Sanskrit: Vidara-Visvasraka. Telugu: Nagajemudu.

Trade name: Prickly pear, Snake hood Fig.

Propagation: by stem cuttings.

Medicinal uses: The plant is used for temporary birth control. The baked fruit is said to be given for whooping cough. The plant is also used in Gonorrhoea and Opthalmia.

Oroxylum indicum (L.) Vent.

Syn: *Bignonia indica* L.

Family: Bignoniaceae.

A small tree, up to 8 m tall. Leaves 2–3-pinnate, with opposite, pinnae, leaflets 2–4 pairs, ovate or elliptic. Flowers pinkish purple in large capsule, brown, flat, woody. Seeds many, winged, thin.

Common in north coastal districts.

Fl. & Fr.: July–March.

Sanskrit: Syonakah, Tuntukah. Telugu: Dundilum, Pampana.

Propagation: by seeds.

Medicinal uses: Roots are used to treat rheumatism, diarrhoea and dysentery. Seeds are used as a purgative. The leaves made into a decoction is given in stomachache and rheumatism and are used externally for enlarged spleen. Tender fruits are refreshing and stomachic.

Ougenia oojeinensis (Roxb.) Hochr.

Syn: *Dalbergia oojeinensis* Roxb.
Ougeinia dalbergioides Benth.

Family: Fabaceae (Papilionaceae.)

Moderate sized deciduous tree, bark black or dark brown. Leaves pinnately 3-foliolate, leaflets broadly ovate or suborbicular, coriaceous. Flowers white or rose-coloured, in short fascicled racemes from the nodes of old branches. Fruit 2–5-jointed pod, flat.

Occasional in north coastal and Telangana districts.

Fl. & Fr.: February–May.

Sanskrit: Tinisa, Syandara. Telugu: Tella moduga, Vandanam.

Trade name: Sandan.

Propagation: by seeds.

Medicinal uses: Bark has stimulant and astringent properties and is used in burning sensation, diarrhoea, dysentery, leprosy, leucoderma, gonorrhoea and diabetes.

Oxalis corniculata L.

Family: Oxalidaceae

Annual or perennial herbs, branchlets creeping, rooting at nodes. Leaves digitately 3-foliolate, leaflets obcordate. Flowers yellow, sub umbellate, axillary. Fruit capsule, oblong, beaked, seeds brown.

Common in moist localities and shade.

Fl. & Fr.: July–December.

Sanskrit: Cangeri. Telugu: Pulichitaku, Anbotikura.

Trade name: Indian sorrel

Medicinal uses: The plant possesses astringent, vermifuge, emmenagogue and antiseptic properties. Fresh juice of the plant cures dyspepsia, piles, anaemia and tympanitis. The leaves of the plant are considered cooling, refrigenant, stomachic, antiscorbutic and appentizing. They are used in fevers, dysentery, scurvy and biliousness and for removing corns, warts and other excrescences of the skin. An infusion of leaves is used to remove opacities of the carnea, it is dropped into sore eyes for itching lids.

Pandanus fascicularis Lam.

Syn: *Pandanus odoratissimus* L.f.
Pandanus tectorius auct.

Family: Pandanaceae.

A much branched shrub or tree, stem supported by prop roots. Leaves linear-ensiform, armed with prickles on margins and dorsal midrib. Flowers unisexual, male flowers white in axillary spikes, female in a solitary terminal cephalium. Fruit an oblong syncarpium.

Occasional in marshy localities.

Fl. & Fr.: July–December.
Sanskrit: Ketaki. Telugu: Mogali, Ketaki.
Trade name: Fragrant screw pine
Propagation: by offsets or suckers.

Medicinal uses: The leaves are said to be valuable in leprosy, small pox, scabies, and the deseases of heart and brain. The anthers of the male flowers are given in ear ache, headache and diseases of blood.

Papaver somniferum L.

Family: Papaveraceae.

Annual erect herb, 30–150 cm tall, latex milky. Leaves entire, glabrous. Flowers solitary, violet-white or red. Fruit round capsule with numerous white seeds.

Rarely cultivated. The cultivation and extraction and sale of opium is banned.

Sanskrit: Ahiphenam, Aphukam. Telugu: Nallamandu (Narcotic), Gasagasalu (seeds).

Trade name: Opium poppy, white poppy.
Propagation: by seeds.

Medicinal uses: Opium is the thickened latex collected from the out side of immature poppy capsules that have had incisions made in the fruit capsules. The unripe capsules suitable for the production of opium are trimmed. Subsequent to drying, the processed latex is scraped off and formed into pieces of varying size. The obtained material is referred to as raw opium and is also the basic substance used for the production of heroin.

Latex contain Isoquinoline alkaloids (20–30%). Chief alkaloids are morphine (3–23%), narcotine (2–10%), codeine (0.2–3.5%), papaverine (0.5–3%) and thebaine (0.2–1%).

The main alkaloid is morphine, which is a strong analgesic, even in small doses, causes euphoria, sedation then narcotic sleep. Opium alone or along with quinine salts, is used in cases of delirium, spasmodic coughing or spasms. Opium is used most frequently as a sedative and /or analgesic.

Pavetta indica L.

Family: Rubiaceae.

Small trees, up to 3 m tall. Leaves chartaceous, oblong-obovate lanceolate. Flowers white in terminal or axillary, trichotomously branched, corymbose cymes. Fruit berry, globose, black when ripe.

Common in forests.

Fl. & Fr.: May–November.
Sanskrit: Papata. Telugu: Papidi.

Trade name: White pavetta.
Propagation: by seeds.

Medicinal uses: The roots are said to possess purgative, aperient, diuretic and tonic properties and are prescribed in visceral obstructions, jaundice, headache, urinary diseases and dropsical affections. A decoction of the leaves is used as a lotion for ulcerated nose and for haemorrhoids.

Pedalium murex L.

Family: Pedaliaceae.

Annual much branched herb, stem herbaceous. Leaves opposite, broad ovate. Flowers purple or pink with yellow throat, axillary, solitary. Fruit hard, indehiscent with two prominent hooks.
Common in waste lands.
Fl. & Fr.: April–September.
Sanskrit: Gokshura. Telugu: Enugu palleru, Pedda palleru.
Propagation: by seeds.

Medicinal uses: A viscid mucilage, resembling that of gum arabic, seperates out by simple agitation of young twigs, leaves fruits or seeds in water or milk. The mucilagenous infusion so formed is credited with demulcent, diuretic and tonic properties, and used in the diseases of the urino-genital system, such as gonorrhoea, dysuria etc. The leaves are applied to ulcers and a decoction of the roots is said to be antibilious.

Pergularia daemia (Forssk.) Chiov.

Syn: *Pergularia extensa* (Jacq.) N.E.Br.
Cynanchum extensum Jacq.
Daemia extensa (Jacq.) R.Br.

Family: Asclepiadaceae.

Perennial, twining, semiwoody, milky weed with foetid, small stem clothed with spreading hairs. Leaves broadly ovate or suborbicular. Flowers greenish yellow in long peduncled fascicles. Fruit paired follicles, curved, softly echinate, seeds with silky coma.
Common in bushes and hedges.
Fl. & Fr.: June–January.
Sanskrit: Karntakah, Uttamarani. Telugu: Gurthichettu, Jithupaku.
Propagation: by seeds.

Medicinal uses: The entire plant is used in medicine. The juice of the leaves is given in catarrhal affection and infantile diarrhoea. It forms a constituent of a purgative preperation given in rheumatism and amenorrhoea. It is used as a uterine tonic, expectorant and emetic. It is useful in curing gynaecological conditions such as excessive bleeding etc.

Phaseolus vulgaris L.

Syn: *Phaseolus nanus* L.

Family: Fabaceae (Papilionaceae)

Annual twining or suberect herbs. Leaves trifoliolate. Flowers white to violet-purple. Fruit pod, straight or slightly curved.

Cultivated for the fruit vegetable.

Telugu: Chikkudu.

Trade name: French bean, Kidney bean.

Propagation: by seeds.

Medicinal uses: Ripe dried pods and the beans are used for curing infections of the urinary tract, kidney and bladder stones. It is also used as a diuretic and antidiabetic.

Phyla nodiflora (L.) Greene

Syn: *Lippia nodiflora* (L.) A.Rich.
Verbena nodiflora L.

Family: Verbenaceae.

Creeping much branched herbs. Leaves obovate or spathulate. Flowers pale pink in axillary dense globose heads. Fruit globose capsule, seeds minute.

Common on river banks, margins of ponds, ditches, canals and moist places.

Fl. & Fr.: Through out the year.

Sanskrit: Jalapippali. Telugu: Bokkenaku.

Trade name: Purple Lippia.

Propagation: by Runners.

Medicinal uses: The plant is said to possess cooling, diuretic and febrifuge properties. It is used in ischury, stoppage of bowels and pain in knee joints. A paste or poultice prepared from the fresh plant is applied as sappurent for boils, swollen cervical glands and chronic indolent ulcers.

Phyllanthus amarus Schum. & Thonn.

Syn: *Phyllanthus niruri* auct. non L.

Family: Euphorbiaceae.

Annual erect herb, up to 50 cm tall, branchlets slender. Leaves elliptic-obovate or oblong. Flowers axillary, unisexual, male flowers solitary in lower axils. Fruit capsule, depressed globose.

Common weed of open waste lands and in cultvated fields.
Fl.& Fr.: Through out the year.
Sanskrit: Bahupatra. Telugu: Nela usiri.
Trade name: Bhuiamla, Kirinelli.
Propagation: by seeds,

Medicinal uses: Whole plant is used to cure jaundice (Viral Hepatites). The plant is bitter in taste and is used as drug. The decoction of the herb is alternative, febrifuge astringent, stomachic and diuretic. It improves appetite quickly. It is also given in dyspepsia, colic, diarrhoea, dropsical affections and other urino-genital diseases.

Active principle: It contains bitter substances phyllanthin (0.35%) and a non bitter substance called hypophyllanthin (0.05%).

Phyllanthus emblica L.

Syn: *Emblica officinalis* Gaertn.

Family: Euphorbiaceae.

A small deciduous tree, up to 10 m tall, branchlets terete. Leaves oblong. Flowers unisexual, in axillary fascicles, male flowers many in upper axils, female flowers few in lower axils. Fruit drupe, depressed globose, fleshy, seeds trigonous.
Common in dry deciduous forests.
Fl. & Fr.: October–December.
Sanskrit: Amlaki. Telugu: Usiri.
Trade name: Indian gooseberry, Indian mylobalan.
Propagation: by seeds.

Medicinal uses: The fruits of *Phyllanthus emblica* are one of the three constituents of the well known Indian preparation Triphala, the other two being *Terminalia bellirica* and *Terminalia chebula.* Triphala is used as a laxative and in treatment of enlarged liver, piles, stomach complaints, pain in eyes etc. Fruits are rich source of vitamin C.

Piper betle L.

Family: Piperaceae.

A perennial climber. Leaves alternate, ovate or ovate-cordate, petiolate, glabrous. Flowers unisexual, in catkins, male and female plants separate in different plants. Fruit sparingly produced, immersed in fleshy spike.
Cultivated all over the state for its leaves which are used as masticatory.
Sanskrit: Tambula, Nagavalli. Telugu: Thamalapaku.

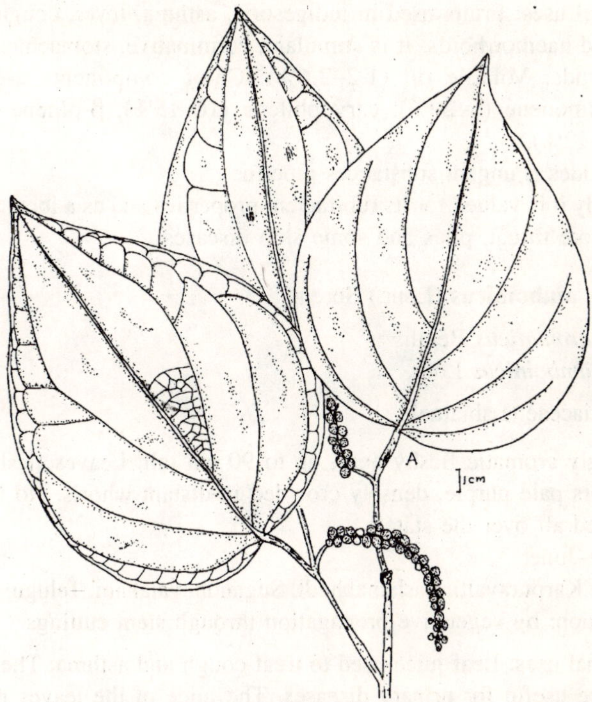

Fig. 92: *Piper nigrum* L.
A. Twig.

Trade name: Betel leaf.
Propagation: stem cuttings.

Medicinal uses: As a masticatory betel leaf is aromatic, digestive, stimulant and carminative. Medicinally it is useful in catarrhal and pulmonary affections. Oil of betel leaf has been used in the treatment of various respiratory catarrhs, and as a local application either by gargle or by inhalation, in diphtheria.

Piper nigrum L.

Family: Piperaceae.

A stout, glabrous climber, stems terete. Leaves broadly ovate. Flowers bisexual in axillary spikes. Fruit black when mature.
Occasional in Nallamalais, cultivated in North coastal districts.
Fl. & Fr.: June–April.
Sanskrit: Maricha. Telugu: Miriyalu.
Trade name: Black pepper, common pepper.
Propagation: Stem cuttings.

Medicinal uses: Fruits used in indigestion, asthma, fever, cough. Arthritis, dysentery and haemorrhoids. It is stimulant, carminative, stomachic.

Compounds: Volatile oil (1.2–2.6%). Chief components are sabinene (15–25%), limonene (5–20%), caryophllene (10–15%), β-pinene (10–12%). α-pinene (8–12%).

Acid amides (Pungent substances-piperine).

Externally it is valued for its rubifacient properties and as a local application for relaxed sore throat, piles and some skin diseases.

Plectranthus amboinicus (Lour.) Spreng.

Syn: *Coleus aromaticus* Benth.
Coleus amboinicus Lour.

Family: Lamiaceae (Labiatae)

A strongly aromatic fleshy herb, up to 90 cm tall. Leaves fleshy, broadly ovate. Flowers pale purple, densely crowded in distant whorls. No fruitng.

Cultivated all over the state.

Fl.: May–June.

Sanskrit: Karpuravalli, Pashanabhedi, Sugandhavalakam. Telugu: Vamu aku.

Propagation: by vegetative propagation through stem cuttings.

Medicinal uses: Leaf juice used to treat cough and asthma. The leaves are reported to be useful for urinary diseases. The juice of the leaves mixed with sugar is a powerful aromatic carminative. It is employed in dyspepsia. Decoction of leaves is used for treating chronic cough and asthma.

Plectranthus barbatus Andr.

Syn: *Coleus barbatus* (Andr.) Benth.
Coleus forskohlii (Willd.) Briq.
Plectranthus forskohlii Willd.

Family: Lamiaceae (Labiatae).

Aromatic herb with thick fleshy root stock, stems ascending, slightly branched. Leaves ovate or obovate. Flowers bluish with red glands in spiciform racemes. Fruit nutlets, globose, brown.

Occasional on laterite and rocky soils in Chittor, Cuddapah, Mahabubnagar and north coastal districts. Now extensively cultivated all over Andhra pradesh.

Fl. & Fr.: August–March.

Sanskrit: Gandira. Telugu: Pashana bhedi.

Trade name: Coleus.

Medicinal uses: Roots contain Forskolin which is used in cough, asthma, carminative. Roots are also used in treating anaemia, inflammation, flatulence dropsy, fever, dysentery, gonorrhoea, epilepsy, abdominal colic, nervous disorders, anaemia, improves vitality, improves appetite, facilitate digestion.

Active principle: Alkaloid Forskolin present in roots.

Plumbago indica L.

Syn: *Plumbago rosea* L.

Family: Plumbaginaceae

A perennial shrub. Leaves broadly ovate or elliptic. Flowers red in terminal and axillary racemes. Fruit a capsule.

Cultivated occasionally.

Sanskrit: Chitrakah. Dahgnah. Telugu: Errachitramulamu.

Trade name: Rosy Flower Lead wort, Rakta chitraka.

Propagation: by offsets or cuttings.

Medicinal uses: Plumbagin acts as a powerful irritant and has antiseptic properties. Plumbagin in small doses has a stimulant action on central Nervous system, on plain muscles and on secretion of sweat, urine and bile. With large doses plumbagin causes paralysis leading to death.

Root is used in the treatment of rheumatism, paralytic affections, colic, inflammations, cough, bronchitis, chronic and intermittant fever, leprosy, leucoderma, ring worm, scabies, anaemia and syphilis.

Active principle: Root bark contain an orange yellow pigment named plumbagin.

Plumbago zeylanica L.

Family: Plumbaginaceae.

Perennial, rambling, under shrub, branchlets striate. Leaves ovate. Flowers white in terminal spikes. Fruit globose capsule.

Occasional in waste lands, hedges and in forests.

Fl. & Fr.: September–December.

Sanskrit: Chitraka, Agni. Telugu: Tellachitramulamu.

Trade name: White flowered Leadwort.

Propagtion: by seeds, roots.

Medicinal uses: Roots are used in the treatment of paralytic affections, ulcers, leprosy, enlarged spleen, piles, skin diseases and influenza. Roots are poisonous and cause abortion. The effect of plumbagin present in roots has been mentioned earlier.

Active principle: Plumbagin in roots.

Plumeria alba L.

Family: Apocynaceae.

A small tree, up to 7 m tall. Leaves obovate or oblanceolate. Flowers white, fragrant in corymbose fascicles. Fruit paired follicles.

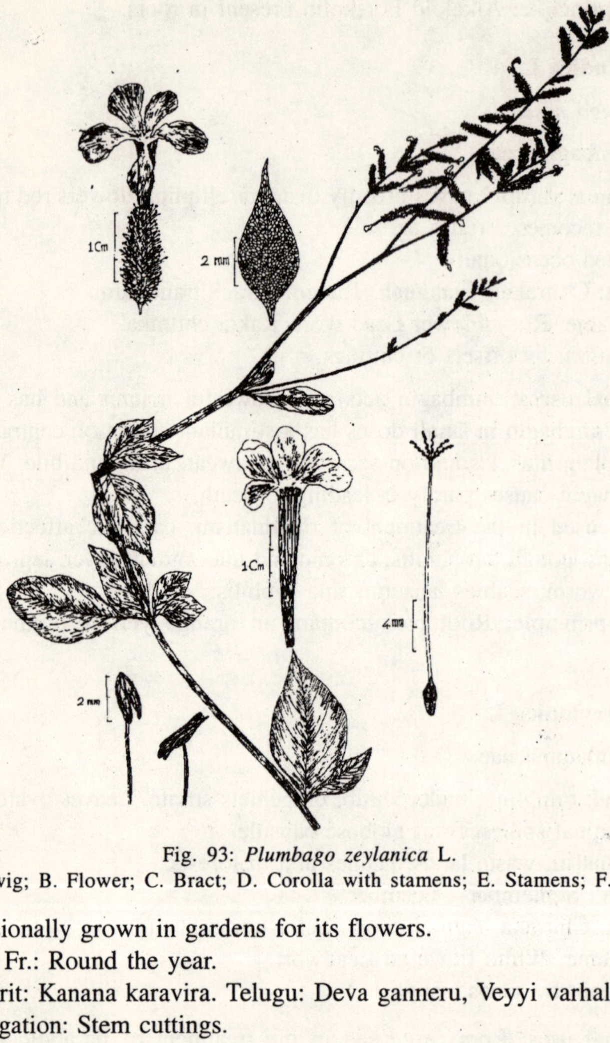

Fig. 93: *Plumbago zeylanica* L.
A. Twig; B. Flower; C. Bract; D. Corolla with stamens; E. Stamens; F. Pistil.

Occasionally grown in gardens for its flowers.

Fl. & Fr.: Round the year.

Sanskrit: Kanana karavira. Telugu: Deva ganneru, Veyyi varhalu.

Propagation: Stem cuttings.

Medicinal uses: The latex is applied to ulcers, herps and scabies and the seeds are said to possess haemostatic properties.

The root is a violent cathartic.

Plumeria rubra L.

Family: Apocynaceae.

Deciduous, profusely branched tree, up to 10 m tall. Leaves broadly obovate-spathulate. Flowers cream with yellow centre. Fruit paired follicles, scarce.

Often planted in temples.

Fl.: April–June Fr.: Scarce.
Sanskrit: Ksiracampaka. Telugu: Devaganneru.
Trade name: Pogado tree.
Propagation: by stem cuttings.

Medicinal uses: Decoction of the bark has been used for the treatment of venereal diseases. It is also used in the indigenous system of medicine for the treatment of rheumatism, diarrhoea, blenorrhoea and leprosy. The flowers are used in pectoral syrups.

Polyalthia longifolia (Sonner) Thw.

Syn: *Uvaria longifolia* Sonner.

Family: Annonaceae

Tall evergreen trees. Leaves narrowly lanceolate, flowers yellowish-green in axillary or terminal fascicles or umbels. Fruit in a cluster of one-seeded berries.
Commonly planted in gardens, avenues and parks.
Fl.: March–April, Fr.: May–September.
Telugu: Ashoka.
Trade name: Mastor cemetery Tree.
Propagation: by seeds.

Medicinal uses: The bark is used as febrifuge. The aqueous extract of bark stimulates the ileum and uterus. The extract lowers blood pressure, stimulates respiration and depresses the heart.

Polygonum chinense L.

Family: Polygonaceae.

Straggling shrub. Leaves broadly elliptic-oblong. Flowers in terminal corymbose panicle. Fruit nutlet, trigonous.
Common along margins of streams.
Fl. & Fr.: September–January.
Propagation: by stem cuttings.

Medicinal uses: The plant is used as a tonic antiscorbutic and vulnerary.

Pongamia pinnata (L.) Pierre

Syn: *Pongamia glabra* Vent.
Cytisus pinnatus L.
Derris indica (Lam.) Bennet.
Galedupa indica Lam.

Family: Leguminosae (Fabaceae)

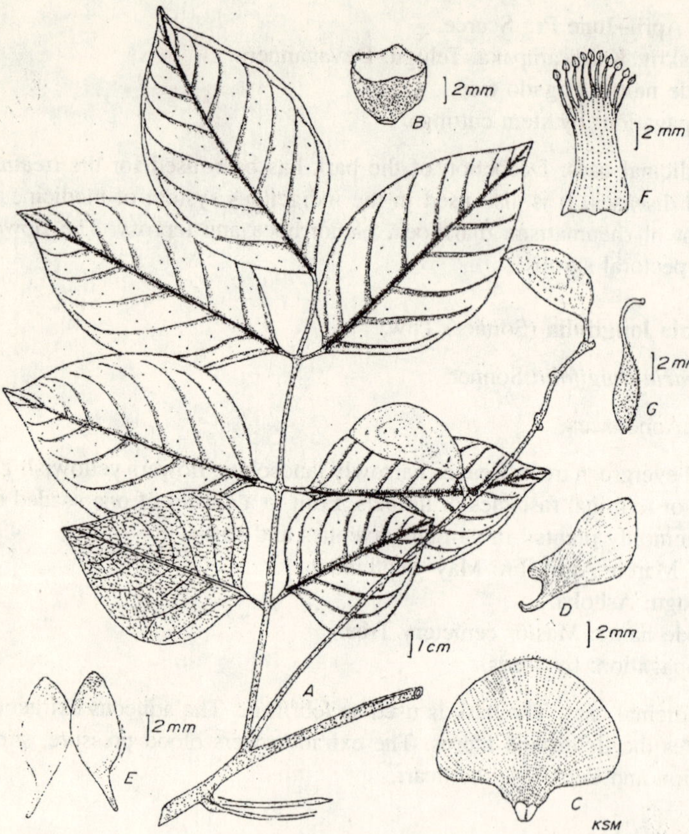

Fig. 94: *Pongamia pinnata* (L.) Pierre
A. Twig; B. Calyx; C. Standard petal; D. Wing petal; E. Keel petal; F. Staminal column; G. Pistil.

Small tree, up to 15 m tall. Leaves imparipinnate, leaflets 5–9, opposite, thin coriaceous, ovate or ovate-oblong. Flowers pinkish-white in axillary racemes. Pod obliquely-oblong or elliptic, 1 or rarely 2-seeded, seed reddish brown.

Common along river banks and in plains. Planted as avenue tree.

Fl.: May–July, Fr.: December–May.

Sanskrit: Karanjah. Telugu: Kanuga chettu.

Trade name: Indian Beech.

Medicinal uses: The juice of this plant is used to treat diarrhoea, cough, dyspepsia, flatulence, leprosy, gonorrhoea. The juice of roots is used to clean teeth, strengthening gums, gonorrhoea and cleansing foul ulcers. A decoction of dried flowers is given in diabetes; seeds yield an oil used in cutaneous diseases like scabies, herpes and leucoderma.

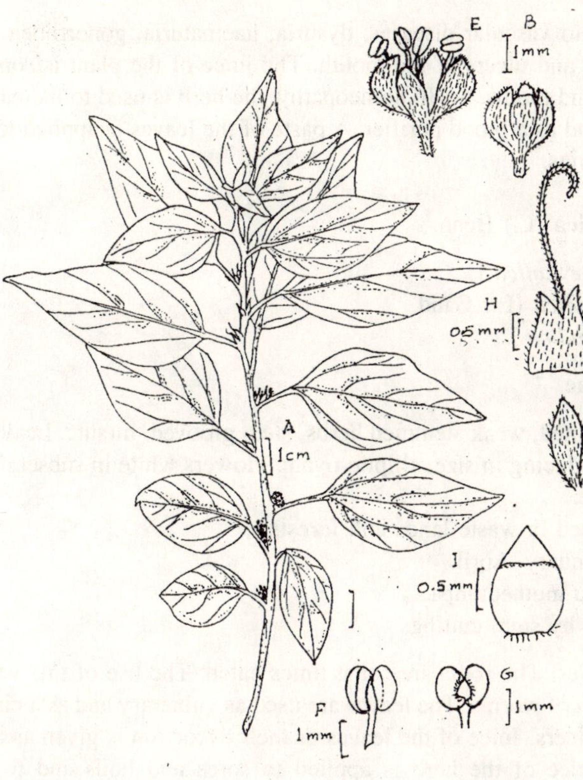

Fig. 95: *Pouzolzia zeylanica* (L.) Bennett
A. Twig, B. Flowers bud; C. & D. Bracteole dorsal and ventral view; E. Male flower; F. & G. Anthers; H. Pistil; I. Achene.

Portulaca oleracea L.

Family: Portulacaceae

Annual prostrate, succulent, glabrous herbs. Leaves linear or obovate-oblong, fleshy. Flowers yellow in terminal clusters. Fruit ovoid capsule. Seeds many, black.

Common weed in cultivated fields and wastelands.

Fl.& Fr.: Round the year.

Sanskrit: Ghotika, Lonika. Telugu: Pedda payali kura, pappukura.

Trade name: Common purslane.

Propagation: by stem cuttings.

Medicinal uses: The herb is considered to possess refrigerant, vulnerary, antiscorbutic, aperient and diuretic properties; the diuretic action is probably due to the presence of high percentage of potassium salts. It is useful in scurvy and in the diseases of liver, spleen, kidney and bladder. It is also prescribed in the treatment of cardio vascular diseases, dysuria, haematuria, gonorrhoea, dysentery, sore nipples and ulcers of the mouth. The juice of the plant is sometimes used in earache and toothache. In homeopathy, the herb is used to increase flow of gastric juice and as a blood purifier. A paste of the leaves is applied to burns, scalds and swellings.

Pouzozia zeylanica (L.) Benn.

Syn: *Parietaria zeylanica* L.
Pouzolzia indica (L.) Gaud.
Parietaria indica L.

Family: Urticaceae.

Perennial, rooted, weak stemmed herbs, stem grooved, hirsute. Leaves variable, gradually reducing in size, elliptic-ovate. Flowers white in subsessile clusters. Fruit achene.

Common weed in waste lands and forests.
Fl. & Fr.: January–April.
Telugu: Eddu muthedumpa.
Propagation: by stem cuttings.

Medicinal uses: The roots are some times eaten. The use of this vegetable leads to expulsion of worms. The leaves are used as vulnerary and as a cicarizant for gangrenous ulcers. Juice of the leaves or their decoction is given as a galactogogue. A poultice of the herb is applied to sores and boils and to relieve stomach ache.

Premna corymbosa (Burm.f.) Rottl.

Syn: *Cornutia corymbosa* Burm.f.

Family: Verbenaceae.

Small trees, spiny. Leaves elliptic-ovate. Flowers greenish white in terminal corymbs. Fruit drupe, black when ripe.

Rare in forests.
Fl. & Fr.: January–September.
Sanskrit: Agnimatha. Telugu: Padmaka, Agnimatha.
Trade name: Fire Brand Teak of Deccan.
Propagation: by stem cuttings, seeds.

Medicinal uses: The roots form a constituent of the well-known Ayurvedic medicine, Dasamula, used as a cure for obstinate fevers. They are prescribed as a laxative, stomachic and cordiac tonic.

Active Principle: Alkaloids premnine, ganiarine and ganicarine.

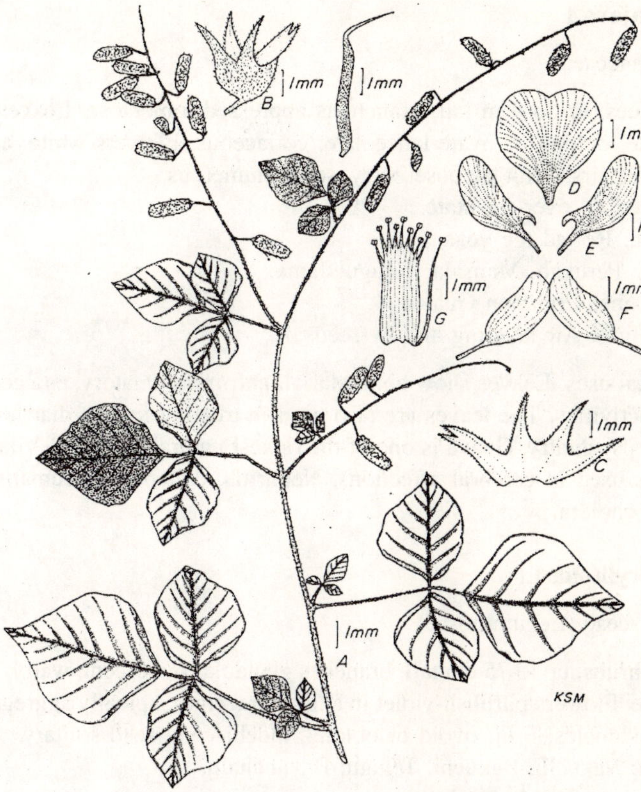

Fig. 96: *Pseudarthria viscida* (L.) Wight & Arn.
A. Twig; B. Calyx; C. Calyx split open; D. Standard petal; E. Wing petals; F. Keel petals; G. Staminal column; H. Pistil.

Pseudarthria viscida (L.) Wight & Arn.

Syn: *Hedysarum viscidum* L.

Family: Leguminosae (Fabaceae)

Diffuse or semierect undershrub, branchlets densely villows, viscid. Leaves pinnately 3-foliolate, terminal ovate-rhomboid, twice as long as laterals, lateral obliquely ovate. Flowers pink in pairs or clusters in axillary and terminal, lax racemes. Fruit linear pod, flat, compressed, hooked, pubescent.

Common in damp forest undergrowth.

Fl. & Fr.: August–January.

Sanskrit: Salaparini. Telugu: Muyyakuponna.

Medicinal uses: The roots of plant possess an astringent taste. They are used in the form of decoction or powder for biliousness, rheumatism, excessive heat and fever, diarrhoea, asthma, heart diseases, worms and piles.

Psidium guajava L.

Family: Myrtaceae.

Small trees, up to 9 m tall, branchlets appressed tomentose. Leaves opposite-decussate, oblong or ovate-lanceolate, coriaceous. Flowers white, axillary, solitary or in pairs. Fruit globose berry, seeds numerous.

Cultivated all over the state.

Fl. & Fr.: Round the year.

Sanskrit: Perukah, Mansala. Telugu: Jama.

Trade name: Common Guava.

Propagation: Air layering and by seeds.

Medicinal uses: Leaves show antimalarial, anti-inflammatory, analgesic and antipyretic activities. The leaves are also used in treating fevers, diarrhoea and as a tonic in psychiatry. Guava is one of the richest natural source of Vitamin C. Root bark is used in cerebral affections, Nephritis, epilepsy, Rheumatism and Vomiting in cholera.

Psoralea corylifolia L.

Family: Fabaceae (Leguminosae)

Under shrubs, up to 75 cm tall, branches glandular pubescent, warty. Leaves simple, ovate. Flowers purplish-violet in axillary racemes, apically aggregated on elongated peduncles. Fruit ovoid or oblong, indehiscent, seed solitary.

Sanskrit: Vakuchi, Bakuchi. Telugu: Bavanchalu.

Trade name: Babchi seed.

Medicinal uses: A crude drug made from this plant is used in India, since long for the treatment of leucoderma. The seeds contain an essential oil which is very effective against certain bacteria causing skin diseases, seeds are useful in treating ulcers, scabies, leprosy and serves as a good hair tonic. Roots of this plant are reported to be useful in treating carries of teeth and the leaves in treating diarrhoea.

Contains Bakuchiol and psoralen used as Anti-leucoderma and antibacterial.

Pterocarpus marsupium Roxb.

Family: Fabaceae (Leguminosae).

Tall deciduous tree, bark thick grey with vertical cracks. Leaves odd pinnate, 5–7-foliolate, leaflets elliptic-oblong, coriaceous. Flowers yellow in terminal panicles. Fruit winged pod.

Common in forests of most districts.

Fl. & Fr.: March–February.

Sanskrit: Asanahm Bijakah. Telugu: Yegisa.

Fig. 97: *Pterocarpus marsupium* Roxb.
A. Twig; B. Calyx split open; D. Standard petal; E. Wing petals; F. Keel petals; G. Staminal column; H. Pistil; I. Flowering twig.

Trade name: Indian kino Tree.

Propagation: by seeds.

Medicinal uses: Kino is a powerful astringent used for diarrhoea and dysentery. It is locally applied in leucorrhoea and in passive haemorrhages. It is also used for toothache. The bark is used as an astringent and in toothache. The flowers are said to be used in fever. The bruised leaves are considered useful as an external application for boils, sores and skin diseases. An aqueous infusion of the wood is said to be of use in diabetes.

Pterocarpus santalinus L.f.

Family: Leguminosae (Fabaceae).

Tall, deciduous tree, bark deeply cleft, brownish black. Leaves 3-foliolate, leaflets ovate-orbicular or oblong. Flowers axillary in simple or sparingly branched racemes. Fruit round pods with a broad wing.

Endemic in the hills of Cuddapah, Nellore and Chittor districts.

Fl. & Fr.: April–July.

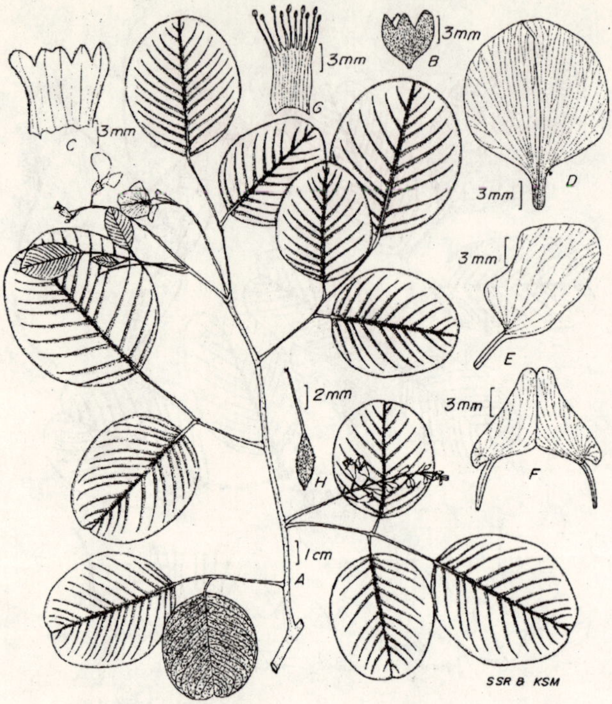

Fig. 98: *Pterocarpus santalinus* L.f.
A. Twig; B. Calyx; C. Calyx split open; D. Standard petal; E. Wing petal; F. Keel petals; G. Staminal column; H. Pistil.

Sanskrit: Raktachandan. Telugu: Yerrachandanamu.

Trade name: Red sanders.

Propagation: by seeds and stumps.

Medicinal uses: Red sanders wood is considered as astringent, tonic and diaphoretic. A paste of the wood is used as a cooling external application for inflammations and headache. It is said to be useful in bilious infections and skin diseases. A decoction of the fruit is used as an astringent tonic in chronic dysentery.

Pterolobium hexapetalum (Roth) Sant. & Wagh

Syn: *Reichardia hexapetala* Roth
Pterolobium lacerans Wight & Arn.
Pterolobium indicum A.Rich.

Family: Caesalpiniaceae.

Straggling shrubs, branches dark pink, prickly, glabrous. Leaves 2-pinnate, prickly, pinnae 6–9 pairs, leaflets 8–10 pairs, ovate-oblong. Flowers white with

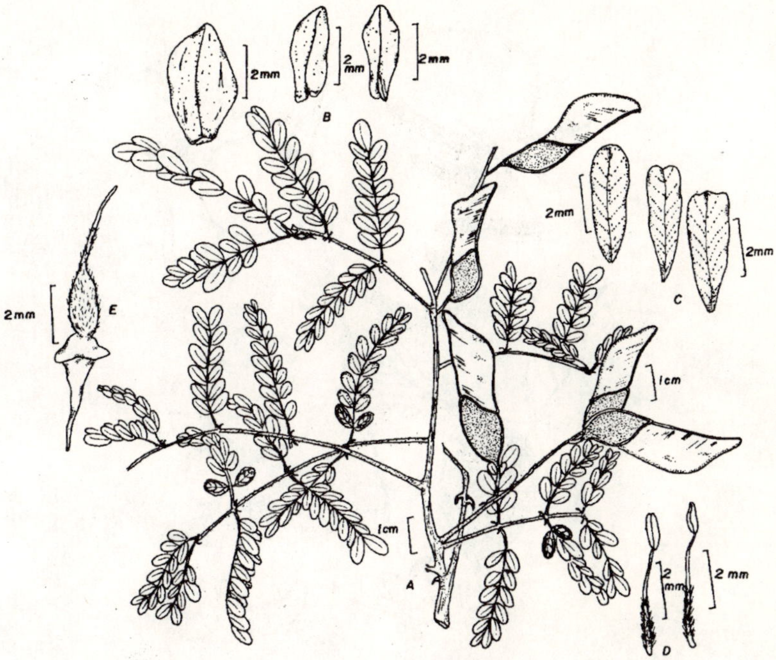

Fig. 99: *Pterolobium hexapetalum* (Roth) Sant. & Wagh
A. Twig; B. Sepals; C. Petals; D. Stamens; E. Pistil.

red tinge in axillary or terminal racemes. Fruit pod, samaroid, oblong, indehiscent, apically winged.

Common in scrub jungles.

Fl. & Fr.: July–December.

Telugu: Korinta, Errachiki.

Trade name: Traveller's terror.

Propagation: by seeds.

Medicinal uses: Young leaves, pods and seed paste are of medicinal use. They are used in diarrhoea, constipation and piles..

Pueraria tuberosa (Roxb. ex Willd.) DC

Syn: *Hedysarum tuberosum* Roxb. ex Willd.

Family: Leguminosae (Fabaceae).

A large woody climber, branchlets appressed velvety. Leaves 3-foliolate, leaflets ovate, laterals oblique, inequilateral, thin coriaceous, glabrous above, appressed white silky beneath. Flowers pale violet in panicles. Fruit pod, linear, seeds 3–6.

Occasional in dry deciduous forests.

Fig. 100: *Pueraria tuberosa* DC.
A. Vegetative twig; B. Flowering twig; C. Fruiting twig; D. Calyx; E. Calyx split open; F. Standard petal; G. Wing petals; H. Keel petals; I. Staminal column; J. Pistil.

Sanskrit: Vidari, Kandapalash. Telugu: Dhari Gummadi.
Trade name: Indian kudzu.
Propagation: by tuberous roots.

Medicinal uses: The tuberous roots are laxative, emetic, diuretic cardiotonic, expectorant and are useful in arthritis, burning sensation, constipation, cardiac debility, intermittent fevers, leprosy, tuberculosis, cough and general debility. Seeds are reportedly good for cutaneous diseases.

Moulali et al (Indian express September,1989) reported that the tubers have aphrodisiac properties. Even the flowers are believed to have aphrodisiac properties.

Punica granatum L.

Family: Punicaceae.

Deciduous, small tree, up to 6 m tall, bark smooth, branchlets often spine tipped. Leaves oblanceolate or oblong. Flowers bright red, solitary or 2–3 together, terminal. Fruit globose berry, crowned by persistent calyx-lobes, interior septate with membranous wall, pulp pellucid, pinkish.

Cultivated for its edible fruit.

Fl. & Fr.: Through out the year.

Sanskrit: Dadimah. Telugu: Danimma.

Trade name: Pomegranate.

Propagation: by seeds.

Medicinal uses: Bark is used for infestation with tape worm and other worms, Bark and rind of the fruit and seeds are used in dysentery and diarrhoea. Fruit peel, stem and root which contains tannins and alkaloids, is anthelmintic and amoeboid.

Active principle: Fruit peel contain 25 to 28% tannins, Bark of stem and root contain 20 to 25% and 0.5% to 1% Alkaloids.

Rauvolfia serpentina (L.) Benth. ex Kurz.

Syn: *Ophioxylon serpentinum* L.

Family: Apocynaceae.

Small erect, perennial undershrub, up to 1 m tall. The root system consists of a prominent tuberous, soft tap root, up to 6 cm in diameter having a corky outer bark. Leaves in whorls of 3, oblanceolate, glabrous. Flowers pink or white in corymbose cymes. Fruit drupe, purplish black when ripe.

Occasional in moist deciduous forests of north coastal districts. Also cultivated, occasionally.

Fl. & Fr.: March–May.

Sanskrit: Sarpagandha. Telugu: Sarpagandha, Pathalagaridi.

Trade name: Serpentine root.

Propagation: The plant grows well in hot humid climate, but prefers shady habitats of forest. It requires very high rainfall ranging from 250 cm to 500 cm and a temperature range of 10–30° C. It can grow in different kinds of soils ranging from red, laterite loam to sandy alluvial loam, but thrives best in well-drained humus-rich clayey soils.

It is propagated by seeds, roots, root stumps and cuttings. As the collection of seeds from wild resources is both laborious and costly, vegetative propagation has been advocated for raising plantation. Root cuttings of 3 to 5 cm long are planted in holes in June and July and are almost completely covered with earth leaving only 1 cm above the surface. The cuttings sprout with in 3 weeks if good moisture is maintained in this period. A success of 50 to 80% has been obtained by various workers in India. About 100 kg of root cuttings are required to plant an hectare of land. The crop propagated by seed is better as yields by this method is larger. About 5 kg of seeds are required per hectare. All seeds are not viable and hence seeds should be used in surplus. Seeds are sown in nusery beds in April–May at a distance of 15–25 cm in rows. The seeds germinate in 4–8 weeks. When the seedlings are three months old they are transplanted into the fields with a spacing of 30 x 45 cm. The field before transplantation is ploughed and 45–50 tons of Farm Yard manure is added. NPK fertilizers at the rate of 80, 40, 40, are applied. The weeds are removed periodically.

Roots (and rhizomes) of the plant constitute the crude drug. As recommended by the Indian Pharmacopoeia the drug is fully formed when the plant is 4 years old. The plants are uprooted carefully during October–November and root parts are seperated. The bark of root is retained in tact. They are thoroughly washed. The roots are sun-dried till they retain a moisture content of 12%. The roots are stored away from light. They are cut into pieces (2–15 cm long and 3–20 mm in diameter). The roots are cylimdrical and almost unbranched. Their colour ranges from light brown to greyish below. The pieces of old roots posses longitudinal ridges while those of young ones wrinkles. The average yield of roots in 1200 kg per hectare.

Medicinal uses: The dried roots of Rauvolfia contain about 0.7–1% alkaloids consisting of atleast 50 compounds. The most important alkaloids are reserpine ($C_{36}H_{40}N_2O_9$), Rescinnamine, Serpentine, Serpentinine, Rawolfine etc. The dried roots of Sarpagandha were used in India to combat hypertension as early as 1000BC. In the early 1950's the preperation attracted the attention of physicians in the west, and a programme for the isolation of active hypotensive agents (agents for lowering blood pressure) was put into action. Of the many alkaloids located the most important was reserpine, which per unit weight is a thousand times as effective as the dried root-preperation itself. Reserpine was first isolated in 1952 by Muller et al. and its molecular structure was deduced in 1953.

Although reserpine is an effective hypotensive agent, it has gained its chief fame as the first of a series of drugs called tranquilizers. It produces drowsiness without coma. The patient is quieted and yet is capable of being aroused at any time. In this reserpine differs from ordinary sedatives.

Reserpine and other traquilizers are useful in the treatment of various psychotic disorders. They are not cures and do not replace psycotherapy. By quietening the patient, however, they may eliminate the more bizarre aspects of this behaviour and make it possible to establish communication between patient and psychiatric so that psychotheray may proceed.

In Ayurveda *Rauvolfia serpentina* roots are used as an antidote for snake bite and stings of insects.

Rauvolfia tetraphylla L.

Syn: *Rauvolfia canescens* L.

Family: Apocynaceae.

Large shrub, 2 m tall. Leaves 4 at each node, elliptic-ovate. Flowers white or pink, small, in terminal 5-7-flowered corymb. Fruit drupe, globose, brownish-purple at maturity.

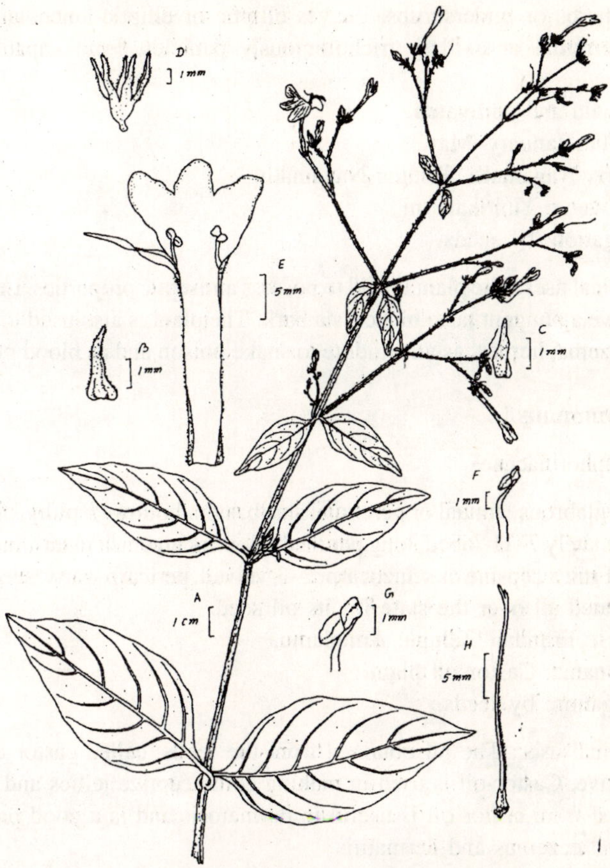

Fig. 101: *Rhinacanthus nasutus* (L.) Kurz.
A. Twig; B. Bract; C. Bracteole; D. Calyx; E. Corolla open; F. Stamen; G. Anther; H. Pistil.

Occasional in Visakhapatnam, Khammam, Cuddapah and Chittor districts.
Fl. & Fr.: September–February.
Telugu: Papataku.
Propagation: by seeds.

Medicinal uses: Roots used to cure Hypertension and skin diseases. Roots of *Rauvolfia tetraphylla* are often used as substitute or adulterant of those of *R. serpentina*.

Rhinacanthus nasutus (L.) Kurz.

Syn: *Justicia nasuta* L.

Family: Acanthaceae.

Erect herbs or undershrubs. Leaves elliptic or elliptic-lanceolate. Flowers white in terminal or axillary trichotomously panicles. Fruit capsule, velvety pubescent.

Both wild and cultivated.

Fl. & Fr.: January–May.

Sanskrit: Nagamalli. Telugu: Nagamalli.

Trade name: Yuthikaparni

Propagation: by seeds

Medicinal uses: The plant is said to possess antiseptic properties. Leaves when chewed have a pungent taste like cassia bark. The plant is also used to cure rheumatism, eczema, herpes, as an antidote to snake poison and as blood purifier.

Ricinus communis L.

Family: Euphorbiaceae.

A tall, glabrous, annual or perennial shrub, stem hollow or pithy, often brittle. Leaves palmately 7–11-lobed, long petioled. Flowers greenish in terminal panicles, unisexual. Fruit a capsule or schizocarp, 2–3-valved, pericarp warty, seeds mottled.

Cultivated all over the state for its oil seed.

Sanskrit: Erandah. Telugu: Amudamu.

Trade name: Castor oil plant.

Propagation: by seeds.

Medicinal uses: The oil obtained from the seeds, called castor oil, is used as a purgative. Castor-oil is used in making contraceptive jellies and creams. A gel prepared from castor oil is useful in dermatosis and is a good protective in occupational eczemas and dermatitis.

A poultice of castor seeds can be applied with beneficial results to gouty and rheumatic swellings. A decoction of the roots of castor plant with carbonate of potash is useful in the treatment of lumbago, rheumatism and sciatica.

A poutice of castor leaves is useful as an external application to boils and swellings.

Castor oil has laxative properties. The laxative principle of castor oil is the ricinolic acid. Ricinolic acid is anti-absorptive and secretogogic.

Chemistry: Castor seeds yield Fatty oil (42 to 55%), Proteic substances (20 to 25%), Lectins (0.1 to 0.7%), Pyrridine alkaloids, Triglycerides and Tecopherols (Vitamin E).

Rivea ornata Choisy

Family: Convolvulaceae.

A climbing or straggling shrub. Leaves orbicular, glabrous above, white silky beneath. Flowers white, fragrant, 3-or more-flowered axillary cyme. Fruit capsule, subglobose, brown.

Rare in forests of Rayalaseema.
Fl. & Fr.: May–November.
Sanskrit: Phanji. Telugu: Boddi Teega.
Propagation: by branch cuttings.

Medicinal uses: The plant is aerial, pungent, and sweetish and is considered cooling and tonic. The juice of the plant is used in the preperation of an ointment for phthriasis and of another for application to piles.

Rostellularia procumbens (L.) Nees

Syn: *Justicia procumbens* L.

Family: Acanthaceae

A diffuse slender herb, up to 30 cm tall. Leaves oblong-elliptic or ovate. Flowers pale pink in terminal, dense, cylindric spikes, bracts obovate, scarious. Fruit capsule, oblong.
In waste lands and open forests.
Fl. & Fr.: September–March.
Propagation: by seeds.

Medicinal uses: It is considered laxative, diaphoretic, diuretic, alterative, expectorant, anthelmintic and febrifuge. The juice of the leaves is squeezed into the eyes in cases of opthalmia. An infusion of the herb is given in asthma, cough, rheumatism, backache, plethora, flatulence and lumbago.

Rotula aquatica Lour.

Syn: *Rhabdia lycioides* C.Martius

Family: Scrophulariaceae

Erect or trailing subshrubs. Leaves fascicled, spathulate. Flowers pink, solitary or 2–3 in axillary clusters. Fruit drupe, subglobose, red when ripe, enveloped by persistent calyx.
Common in river beds.
Fl. & Fr.: February–June.
Sanskrit: Pashanabheda. Telugu: Pashnabheda.
Propagation: by seeds.

Medicinal uses: Roots (Pashanabheda) are valued in Ayurvedic medicine. A decoction of roots is used as diuretic and laxative and also for piles, stones in the bladder and in venerial diseases.
Chemistry: Roots contain allatonin and a sterol.

Rubia cordifolia L.

Syn: *Rubia munjista* Roxb.

Family: Rubiaceae.

A rugose vine or herb, branches 4-angled. Leaves whorled, 4 at the node, cordate. Flowers white in terminal and axillary dichasial cymes. Fruit globose drupe, fleshy.

Rare in moist deciduous forests of North coastal Andhra pradesh and Chittoor districts.

Fl. & Fr.: June–October.

Sanskrit: Manjistha, Kala-Meshika, Yojanavalli. Telugu: Tamravalli. Manjestatiga.

Trade name: Indian Madder.

Propagation: by seeds.

Fig. 102: *Salvadora persica* L.
A. Twig; B. Flower; C. Corolla; D. Stamen; E. T.S. of ovary.

Medicinal uses: Roots are used to treat rheumatoid arthitis, flatulence, diarrhoea, dysentery, leprosy, skin diseases, leucoderma wounds, ulcers, diabetes, slow healing of bróken bones and general debility. Extracts form a constituent of the drug septilin used for rhinosinal infections.

Chemistry: Colouring matter present is a mixture of purpurin and manjistin.

Salvadora persica L. var. **wightiana** (Planch ex Thw.) Verdc.

Family: Salvadoraceae.

Large much branched, evergreen shrub or small tree, up to 10 m tall. Leaves elliptic-oblong. Flowers greenish-yellow in axillary and terminal panicles. Fruit fleshy drupe, spherical, red when ripe.

Commonly planted in Mosques and Muslim burial grounds.

Fl. & Fr.: November–April.

Sanskrit: Pilnh. Telugu: Gogu Gunia, Wargu.

Trade name: Salt Bush Tree, Tooth Brush Tree, Meshwak.

Propagation: by seeds.

Medicinal uses: Fresh bark is used as a vesicant. Root decoction is used against gonorrhoea and vesical catarrh. The extract of the root is said to relieve the pain due to spleen troubles. A decotion of the bark is used as a tonic and stimulant in low fevers and as an emmenagogue. Leaves possess antiscorbutic and astringent properties. A decoction of the leaves is used in asthma and cough, and a poultice made out of them is applied to painful tumors and piles. They are also used as an external application in rheumatism. Fruits possess deobstruent, carminative, diuretical and stomachic properties and are used in biliousness and rheumatism.

Sansevieria roxbughiana Schult. & Schult.f.

Syn: *Sansevieria zeylanica* Roxb.

Family: Agavaceae.

A herb with very short stem or stemless, root stock creeping. Leaves about 8 to 10 in a tuft, rigid, pale green with transverse bands of dark green. Flowers greenish-white tinged with violet in fascicles of 3–6 on long racemes. Fruit globose, fleshy.

Occasional in forests, also cultivated as ornamental.

Fl. & Fr.: June–August.

Sanskrit: Murva Maurvi. Telugu: Niyanda, Sagal.

Trade name: Bow string Hemp.

Propagation: by suckers.

Medicinal uses: Rhizome is used in the form of an electuary for coughs. The juice of tender shoots is administered to children to clear their throats of phlegm.

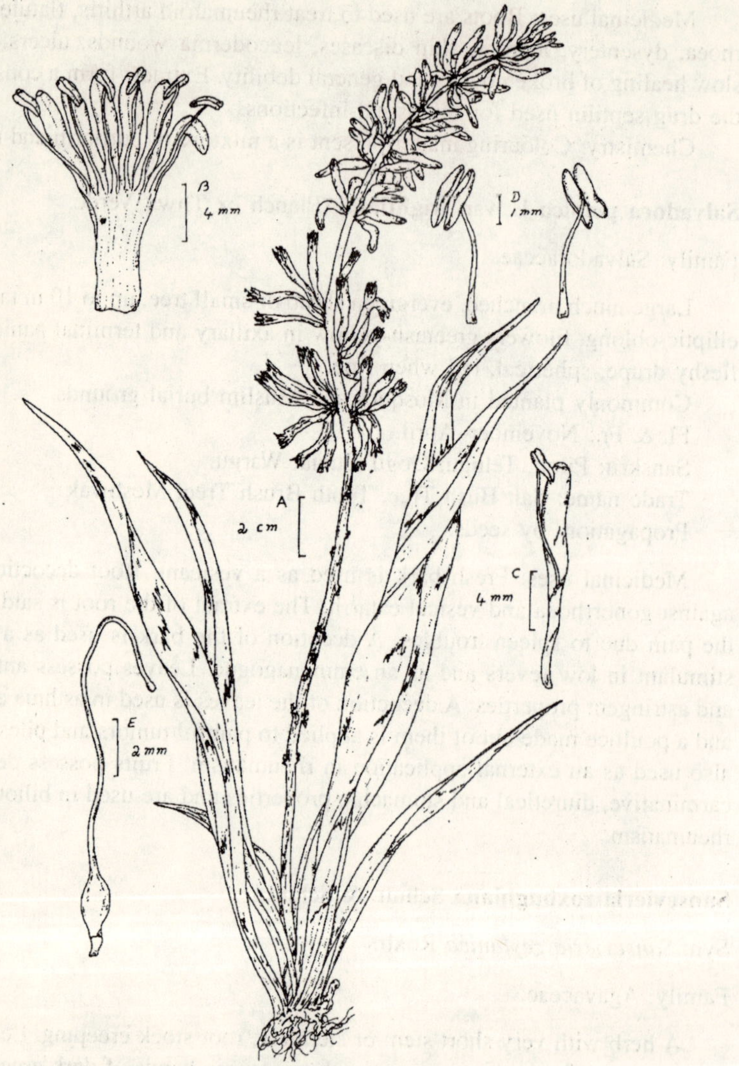

Fig. 103: *Sansevieria roxbughiana* Schult. & Schult. f.
A. Habit; B. Perianth split open with stamens; C. Perianth lobe with stamens; D. Stamens; E. Pistil.

Santalum album L.

Family: Santalaceae.

Small evergreen tree with slender drooping branches. The sapwood white and odorless, the heartwood yellowish brown, strongly scented. Leaves elliptic ovate to lanceolate. Flowers in terminal and axillary paniculate cymes.

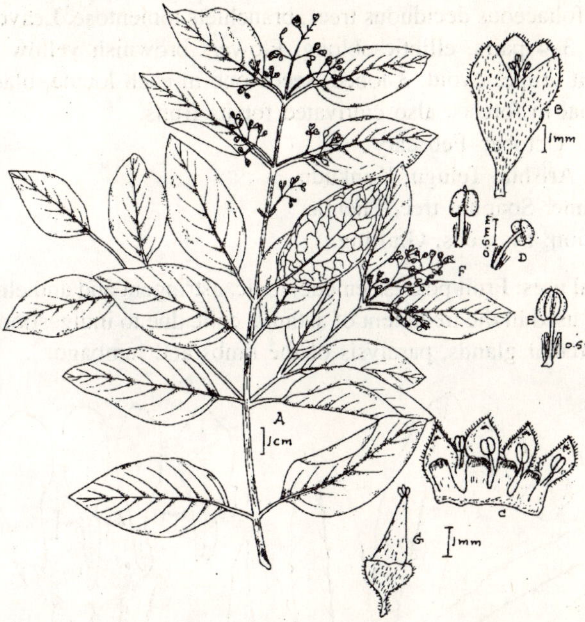

Fig. 104: *Santalum album* L.
A. Twig; B. Flower; C. Corolla split open; D.-F. Stamens; G. Pistil.

Common in forests of Anantapur district, planted elsewhere.
Fl. & Fr. January–October.
Sanskrit: Chandana. Telugu: Chandanamu, Sreegandhamu, Gandhapuchekka.
Trade name: Sandal tree.
Propagation: by seeds.

Medicinal uses: Sandal wood oil extracted from the heart wood of the tree is used for inflammatory conditions of the efferent urinary tract. Sandal wood oil is also used in treatment of dysuria (to promote and facilitate urination), gonorrhoea and cough. The wood, ground up with water into a paste, is applied on local inflammations on forehead in fever and on skin diseases. Oil from the seed is used on skin diseases.

The paste of the wood and oil is used for treating burning sensation, skin diseases, cardiac debility, jaundice, cough, bronchitis, gastric irritability, intermittant fever and general debility. The bark is used for treating malaria. The drug of this plant is used to treat tuberculosis and gall bladder.

Sapindus emarginatus Vahl

Syn: *Sapindus trifoliatus* auct.

Family: Sapindaceae.

Densely foliaceous deciduous trees, branchlets tomentose. Leaves even-pinnate, leaflets 3–4 pairs, elliptic-oblong. Flowers brownish yellow in terminal panicles. Fruit drupe, ovoid, 3-lobed, seeds one in each locule, black.

Occasional in forests, also cultivated for its fruits.

Fl. & Fr.: October–February.

Sanskrit: Arishta. Telugu: Kunkudu.

Trade name: Soapnut tree, Phenila.

Propagation: by seeds, Grafting.

Medicinal uses: Fruit possess emetic, tonic, astringent and anthelmintic properties and are used in the treatment of asthma, colic due to indigestion, diarrhoea, cholera, tubercular glands, paralysis of the limbs and lumbago.

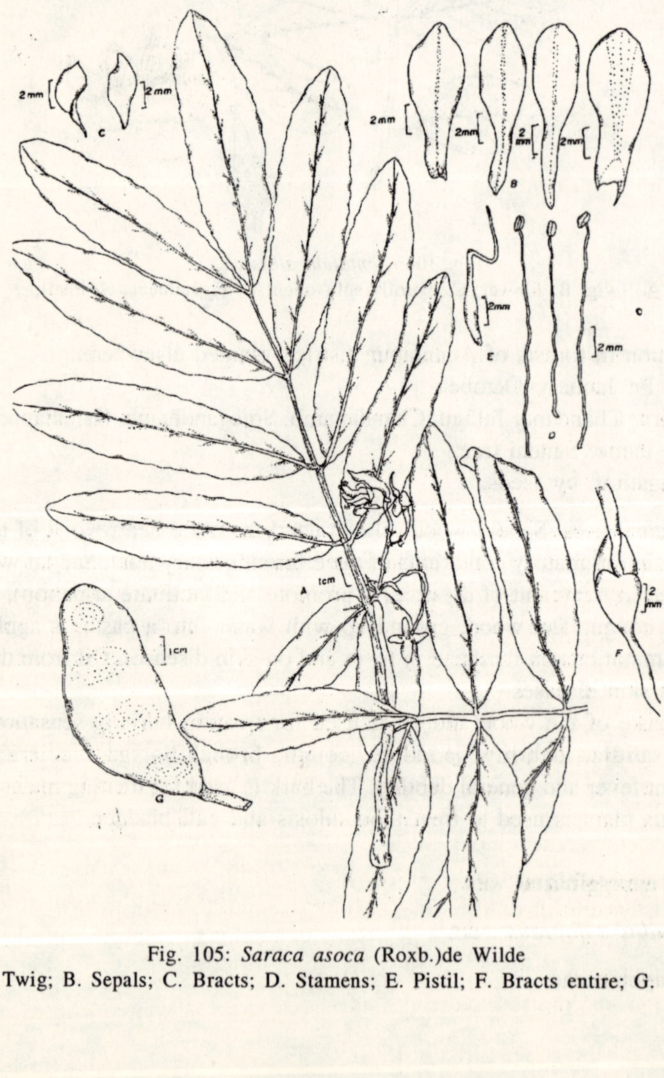

Fig. 105: *Saraca asoca* (Roxb.)de Wilde
A. Twig; B. Sepals; C. Bracts; D. Stamens; E. Pistil; F. Bracts entire; G. Pod.

Saraca asoca (Roxb.) de Wilde

Syn: *Jonesia asoca* Roxb.

Saraca indica auct non L.

Family: Caesalpiniaceae

Ever green tree, up to 5 m tall. Leaves even pinnate, leaflets 4–6 pairs, oblong, coriaceous. Flowers brilliant orange-yellow, pink with age in short dense corymbose panicles. Fruit oblong pod, compressed, flat.

Common in forests of Srikakulam and Visakhapatnam districts.

Fl. & Fr.: All seasons.

Sanskrit: Ashoka. Telugu: Ashoka, Vanjulamu.

Trade name: Ashoka.

Medicinal uses: Dried bark is astringent and is used to treat excessive mensturation, uterine affections, gynaecological problems and as a uterine sedative. Seeds are useful in urinary discharges. Flowers pounded and mixed with water are useful in treatment of haemorrhagic dysentery. Flowers are considered to be an excellent uterine tonic. They are also considered useful in biliousness and syphilis. Flowers pounded in water are used in haemorrhagic dysentery and the dried flowers in diabetes.

Sarcostemma acidum (Roxb.)Voigt

Syn: *Asclepias acida* Roxb.

Sarcostemma brevistigma Wight & Arn.

Family: Asclepiadaceae

Much branched, leafless, climbing, succulent jointed shrub, stem green. Flowers creamy-white, sessile in terminal umbellate cymes.

Abundant especially in stony places and waste lands.

Fl. & Fr.: June–September.

Sanskrit: Somavalli, Soma. Telugu: Kondapala, Palmavashtam.

Trade name: Moon plant, Moon creeper.

Propagation: by stem cuttings, by seeds.

Medicinal uses: The plant is considered emetic, alterative, narcotic and rejuvenating and used in hydrophobia, psychopathy and general debility.

Schrebera swietenioides Roxb.

Family: Liliaceae.

Moderate sized deciduous tree, up to 18 m tall, bark greenish, rough with protuberances. Leaves imparipinnate, leaflets 7–9, elliptic-ovate, coriaceous. Flowers yellowish-white in terminal and axillary 3-chotomously paniculate cymes, covered with brown scales. Fruit pear-shaped loculicidal capsule, very hard and woody, dotted with scabrous specks, 2-valved, seeds winged.

Occasional in moist deciduous forests.

Fl. & Fr.: January–May.

Sanskrit: Muskaka Patali. Telugu: Tondamukkudi, Mokkam.

Trade name: Weaver's Beam tree.

Propagation: by seeds and through stem cuttings.

Medicinal uses: Leaves are used in treating enlargement of spleen and urinary discharges. Roots used in leprosy. Bark is used against boils and burns. The fruits are digestive, purgative and stomachic and is useful in flatulence, anorexia, colic, anemia and diabetes.

Semecarpus ancardium L.f.

Family: Anacardiaceae.

Moderate sized deciduous tree with red blaze exuding resin which blackens on exposure. Leaves very large crowded at the ends of branches. Flowers greenish-white in terminal panicles. Fruit oblong-ovoid drupe, shining black when ripe.

Occasional in dry deciduous forests.

Fl. & Fr.: May–December.

Sanskrit: Bhallataka. Telugu: Jidi, Bhallatakamu.

Trade name: Marking nut, Oriental Cashew.

Propagation: by seeds.

Medicinal uses: The ripe fruit and its oil is used for treating dyspepsia, nervous debility, neurites, rheumatism and impotency. The fruits are used to treat cancer, flatulence, constipation, helminthiasis, leprosy, leucoderma, cardiac diseases, diabetes, tumors, ulcers and general debility.

Senecio tenuifolius Burm.f.

Family: Asteraceae (Compositae).

Annual erect, glabrous herbs, up to 50 cm tall. Leaves pinnatifid or bipinnatifid. Flowers yellow, in terminal heads, heterogamous. Fruit achene, obovate, ribbed, hairy between ribs.

Occasional in forests in Rayalaseema.

Fl. & Fr.: September–December.

Telugu: Katimera chamanthi.

Propagation: by seeds.

Medicinal uses: Leaves are emollient and vulnerary.

Sesamum orientale L.

Syn: *Sesamum indicum* L.

Family: Pedaliaceae.

An annual erect, pubescent herb. Lower leaves lobed or deeply divided, upper ones undivided, linear-oblong. Flowers axillary, solitary, pink or rosy. Fruit capsule.

Cultivated for oil seed.

Fl. & Fr.: Through out the year.

Sanskrit: Tila. Telugu: Nuvvulu.

Trade name: Sesame, Gingelly.

Propagation: by seeds.

Medicinal uses: Sesame seeds are considered emollient, diuretic, lactogogue and a nourishing tonic. They are said to be helpful in piles, a paste of the seeds mixed with butter being used in bleeding piles. A decoction of the seed is said to be an emmenagogue and is also given in cough. The oil is employed as vehicle for injections and also used as emollient and demulcent.

Sesbania grandiflora (L.) Poir.

Syn: *Robinia grandiflora* L.

Family: Fabaceae (Leguminosae).

Small trees, branchlets pubescent. Leaves compound, leaflets 25–30 pairs, oblong or slightly obovate. Flowers cream-coloured in 3–5-flowered racemes. Fruit pod, marigin thick, seeds up to 30.

Cultivated in betel-vine plantations.

Fl. & Fr.: December–March.

Sanskrit: Agastyah, Agathi. Telugu: Avise.

Trade name: Swamp pea, Sesban, Agathi.

Propagation: by seeds.

Medicinal uses: Leaves are aperient, tonic, diuretic and laxative. The juice of the leaves as also that of the flowers is used in nasal catarrh and headache. A poultice made from the leaf juice is applied to bruises. Seeds are used as emmenagogue. The juice of the root is given with honey as an expectorant. A paste of equal quantities of roots of red flowered agathi and stramonium is applied to painful swellings. A paste of the roots in water is applied in rheumatism. Agathi bark is astringent, bitter, tonic and a febrifuge. A decoction of the bark in small doses is taken against diarrhoea and dysentery, but in large doses it is emetic and aperient. Pounded bark is externally applied to cure scabies.

Shorea robusta Gaertn.

Syn: *Vatica robusta* Steud.

Family: Dipterocarpaceae.

Very large trees. Leaves ovate. Flowers pale yellow in terminal or axillary lax panicles. Fruit leathery capsule, winged, wings 6–8 times as long as fruit.

Rare in forests of North coastal districts.
Fl. & Fr.: March–May.
Sanskrit: Salah, Asvakarnah. Telugu: Guggilamu, Sarjakamu.
Trade name: Sal.
Propagation: by seeds.

Medicinal uses: A paste made from fruits is used in diarrhoea. essential oil distilled from resin is used for skin diseases and for ear troubles. The resin is used as an astringent and is given in diarrhoea and dysentery. It is also used as an ingredient of ointments for skin diseases and in ear troubles.

Shorea tumbaggaia Roxb.

Syn: *Vatica tumbaggaia* (Roxb.) Wight & Arn.

Family: Dipterocarpaceae.

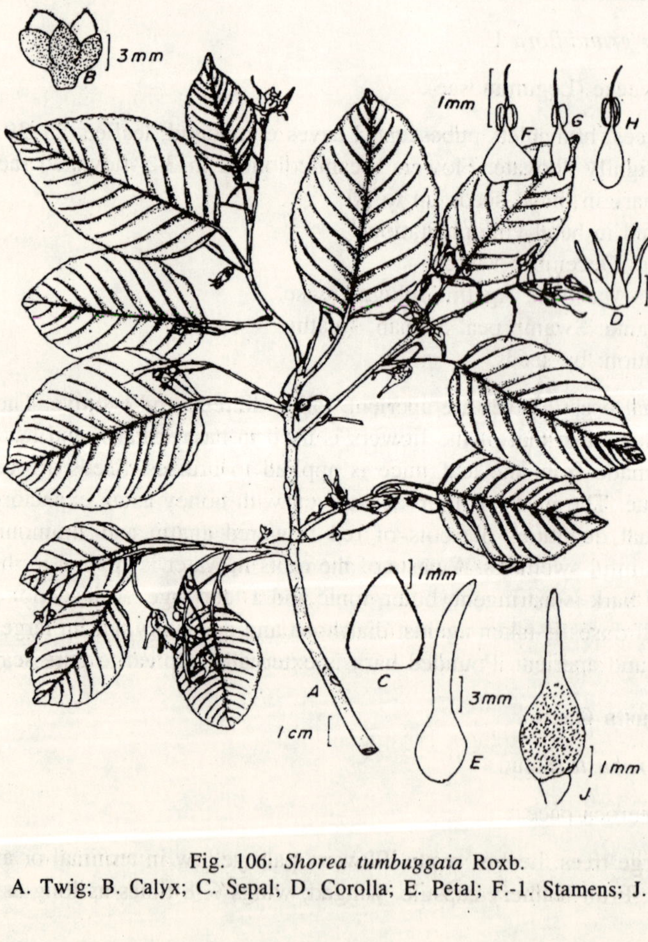

Fig. 106: *Shorea tumbuggaia* Roxb.
A. Twig; B. Calyx; C. Sepal; D. Corolla; E. Petal; F.-1 Stamens; J. Pistil.

Deciduous trees, bark dark brown, longitudinally fissured. Leaves ovate, glabrous, thick coriaceous. Flowers white, fragrant in terminal panicles. Fruit capsule, winged, yellowish white.

Endemic to Seshachalam hills and Veligonda hills.

Fl.& Fr.: April–July.

Telugu: Tamba.

Trade name: Dammar Tree.

Propagation: by seeds.

Medicinal uses: Resin is used as an external stimulent.

Sida cordifolia L.

Family: Malvaceae.

Much branched undershrubs, branchlets densely stellate-tomentose. Leaves cordate or ovate, tomentose. Flowers pale yellow, axillary, solitary. Fruit schizocarp, mericarps 8–10. A weed of waste lands and road sides all over the state.

Fl.& Fr.: August–November.

Sanskrit: Bala, Kathorayashtika. Telugu: Chirubenda, Tellantisa, Suvarnamu.

Trade name: Country Mallow.

Propagation: by seeds.

Medicinal uses: The mucilaginous leaves are used as a demulcent and their infusion is given in fever as refrigerant. They are used against dysentery and ulcers. A decoction of the leaves is said to possess emollient and diuretic properties.

The plant is used as a general tonic and for improving sexual strength. Seeds of the plant are particularly credited with this property.

Seeds are considered useful in gonorrhoea and colic pains.

Chemistry: Alkaloids in the whole plant 0.085%.

Sida rhombifolia L.

Syn: *Sida rhomboidea* Roxb. ex Flem.

Family: Malvaceae.

Erect perennial herb. Leaves ovate-elliptic or rhomboid, stellately hairy above, wooly below. Fruit schizocarp, rugulose, seeds 7–10, ovoid.

Fl.& Fr.: July–December.

Sanskrit: Atibala. Telugu: Pulagamu, Ativala.

Trade name: Barela.

Propagation: by seeds.

Medicinal uses: Stem is employed as demulcent and emollient. It is also used internally in skin diseases and as a diuretic and febrifuge. Roots are used in the tratment of rheumatism and leucorrhoea. The plant is a major ingredient in several Ayurvedic medicines.

Sida spinosa L.

Family: Malvaceae.

Woody, much branched, erect herbs-undershrubs, stems and branches rough with spiny tubercles at the nodes below the petioles. Leaves ovate-elliptic or oblong. Flowers yellow with red tinge, axillary, solitary or 2–5 in clusters. Fruit schizocarp, wrinkled, mericarps 5, seeds ovoid.

Occasional weed of waste places and forests in most districts.

Fl.& fr.: September–March.

Sanskrit: Mahabala. Telugu: Nagabala.

Trade name: Prickly Sida.

Propagation: by seeds.

Medicinal uses: Leaves are reported to possess demulcent and refrigerant properties and useful in cases of gonorrhoea, gleet and scalding urine. Fruits are credited with astringent and cooling properties. Root is used as a tonic and diaphoretic and is given in mild cases of debility and fever. A decoction of it is said to be given as demulcent in irritability of bladder and in gonorrhoea.

Smilax zeylanica L.

Syn: *Smilax macrophylla* Roxb.

Family: Smilacaceae.

A large climber, stem smooth, striate, armed with a few small distant prickles. Leaves alternate, elliptic or ovate-lanceolate, coriaceous, glabrous. Flowers in pedunculate many-flowered umbels, unisexual. Fruit globose berry, red when ripe.

Common in all forests.

Fl.& Fr.: June–September.

Telugu: Konda Danthina, Phirangi Mokka.

Trade name: Rough bind weed.

Medicinal uses: Decoction of roots is given for swellings, abscesses and boils, venereal and skin diseases.

Solanum melongena L.

Family: Solanaceae.

Armed undershrubs, up to 1m tall. Leaves angulate, stellate-pubescent, prickly along the nerves. Flowers violet, in extra axillary cymes. Fruit berry, oblong-globose, fruiting calyx enlarged.

Cultivated for its fruit vegetable.
Sanskrit: Varthaki, Hingoli. Telugu: Vankaya.
Trade name: Brinjal, Egg plant.
Propagation: by seeds.

Medicinal uses: Roots of brinjal plants are credited in the indigenous medicine as antiasthmatic and general stimulant. Leaves are said to possess sialogogue and narcotic properties and are used in cholera, bronchitis, dysuria and asthma. Brinjals are recommended in liver complaints. The seeds are used as a stimulant.

Solanum nigrum L.

Family: Solanaceae.

An annual, erect, much branched, unarmed herb, up to 1m tall. Leaves ovate, glabrous. Flowers white, in extra axillary cymes. Fruit globose berry, glabrous, purplish red when ripe, seeds many, ovate-reniform.
Common weed of waste places, road sides and cultivated fields.
Fl. & fr.: Round the year.
Propagation: by seeds.
Sanskrit: Kakamaci. Telugu: Kamanchi.
Trade name: Black Night-Shade.

Medicinal uses: A freshly prepared extract of the herb is effective in cirrhosis of liver. In traditional medicine, the plant has been used as a febrifuge, antidiarrhoeal, in eye diseases and hydrophobia. The fruits have been reported to possess tonic, diuretic Cathartic and anti-hepatotoxic potential. The herb has antiseptic and antidysenteric properties and is given internally for cardalgia and gripe. The plant is also credited with emollient, diuretic and laxative properties and its decoction is regarded as an antispasmodic and narcotic.

Solanum surattense Burm. f.

Syn: *Solanum xanthocarpum* Schrad.

Family: Solanaceae.

A perennial very prickly, diffuse herb. Leaves pinnatifid, ovate-elliptic. Flowers bluish-purple in 4–5-flowered extra axillary cymes. Fruit globose berry, yellow when ripe.
Most striking weed in plains in waste places and road sides, after harvesting season in fields in all districts.
Fl. & Fr.: Round the year.
Sanskrit: Kantakari. Telugu: Nelamulaka, Vakudu.
Trade name: Yellow-Berried Night shade.

Medicinal uses: Root is an expectorant, forming an ingredient of the well-known Ayurvedic medicine Dasamula. It is employed in cough, asthma and pains in chest.

Stem, flowers and fruits are bitter and carminative. They are prescribed for relief in burning sensation in the feet accompanied by vesicular watery eruptions. Leaves are applied locally to releive pain. Seeds are also used as an expectorant in asthma and cough. The fruits of the plant are considered useful in a number of diseases, such as sore throat bronchitis, muscular pains, fevers etc.

Solanum trilobatum L.

Family: Solanaceae

Armed, climbing undershrub. Leaves ovate-angular, prickly. Flowers violet, in extra-axillary cymes. Fruit globose berry, seeds circular.

Occasional in plains especially on the coast.
Fl. & Fr.: December–March.
Sanskrit: Alarka. Telugu: Vuschinta ush.
Trade name: Climbing brinjal.
Propagation: by seeds.

Medicinal uses: The bitter roots are used for consumption in the form of an electuary, decoction or powder. The berries and flowers are administered for treatment of cough. The decocotion of various parts of the plant is used in chronic bronchitis. The plant is reported to possess antibacterial, antifungal, antimitotic and antitumor activites.

Chemistry: Steroidal alkaloid Solasodine is present in fruit and leaf of the plant.

Sphaeranthus indicus L.

Family: Asteraceae (Compositae).

Annual diffuse or suberect aromatic herb, stems winged. Leaves obovate or spathulate. Flowers pink-purple in sessile, terminal heads. Fruit oblong achene.

Common in dried ponds, ditches and dried rice fields.
Fl. & Fr.: November–April.
Sanskrit: Mundi, Sarvani. Telugu: Bodasaramu.
Trade name: East Indian Globe Thistle.
Propagation: by seeds.

Medicinal uses: The juice of the plant is styptic and said to be useful in liver and gastric disorders. The paste of the herb, made with oil, is applied in itch. The powdered seeds and roots are given as an anthelmintic. A decoction of the root is used in chest pains, cough and bowel complaints. Flowers are credited with alterative, depurative and tonic properties. Leaf juice is boiled with milk and sugar-candy and prescribed for cough.

Spondias pinnata (L.f.)Kurz.

Syn: *Mangifera pinnata* L.f.
Spondias mangifera Willd.
Spondias acuminata Roxb.

Family: Anacardiaceae.

Moderate sized deciduous tree. Leaves imparipinnate, leaflets opposite, 3 pairs, oblong-elliptic. Flowers white with yellow tinge in terminal panicles. Fruit globose-ovoid drupe, seated on a fleshy receptacle, seed thick.

Occasional in dry deciduous forests.

Fl. & Fr.: January–December.

Sanskrit: Amratakah. Telugu: Adavimamidi, Kondamamidi.

Trade name: Wild Mango, Hog Plum.

Propagation: by seeds.

Stachytarpheta jamaicensis (L.) Vahl

Syn: *Verbena jamaicensis* L.
Stachytarpheta indica auct.

Family: Verbenaceae.

Erect woody herbs. Leaves obovate-elliptic. Flowers blue in long slender terminal spikes. Fruit oblong, ribbed.

Common in moist localities.

Fl. & Fr.: August–January.

Trade name: Bastard vervain, Brazilian tree.

Medicinal uses:The herb is used for treating intestinal worms, venereal diseases, ulcers, erysipelas, dropsy and stomach ailments. An infusion of the bark is used against diarrhoea and dysentery. Leaves are said to be used n cordiac troubles and are also rubbed on sprains and bruises.

Chemistry: Plant is said to contain a glucoside, stachytarphine and an alkaloid.

Stemona tuberosa Lour.

Family: Stemonaceae.

Twiner with tuberous rhizome. Leaves ovate. Flowers bisexual, axillary. Fruit ovoid-oblong capsule, 5–8-seeded.

Occasional in north coastal districts and Cuddapah hills.

Telugu: Kanipu teega, Ijedigadda.

Medicinal uses: The tuberous roots show bacteriostatic activity and are used in phthysis and coughs. The drug soothes the respiratory centres without affecting the heart.

Sterculia urens Roxb.

Family: Sterculiaceae.

Soft wooded tree with papery outer bark. Leaves palmately 3-or 5-lobed crowded at the tips of the branchlets, glabrous above, velvety below. Flowers greenish yellow in terminal panicle, glandular pubescent. Fruit of 5 follicles, radiating, pubescent with stinging hairs, seeds oblong, black.

Common in Deciduous forests.
Fl. & Fr.: February–August.
Telugu: Tabsu, Tapsi, Kovila Chettu.
Trade name: Gum karya, Kateera Gum.

Medicinal uses: The tree yields a gum, known as gum karaya. The gum is much used in pharmaceuticsl and medicinal preperations, such as lozenges, emulsions, lotions, sprays and pastes. Another important pharmaceutical application of the gum is in denture-fixative powders. The gum is also used as a laxative. The gum is also reported to be useful for the affections of throat. It is also used in the preperations of the cosmetics.

Stereospermum suaveolens (Roxb.) DC.

Syn: *Bignonia suaveolens* Roxb.

Family: Bignoniaceae

Large deciduous trees. Leaves 1-pinnate, leaflets 5–9, elliptic. Flowers dull crimson in large panicles. Fruit capsule, straight.

Occasional in forests of Visakhapatnam district and Nallamalais.
Fl. & Fr.: September–February.
Sanskrit: Patala, Patali. Telugu: Kokkesa, Patala.
Trade name: Paarul mool.
Propagation: by seeds.

Medicinal uses: Roots are used to treat intermittant fevers, inflammatory affections of the chest, affections of the brain, dropsy and dyspepsia. The root bark is a constituent of the well known Ayurvedic formulation "Dashamoola." Flowers are given with honey to control hiccup.

Strychnos colubrina L.

Family: Loganiaceae

Large climbing shrub, bark smooth, tendril single, bifid. Leaves ovate to elliptic, coriaceous, glabrous, shining. Flowers in small, compound, axillary puberulous cymes on old wood. Fruit crustaceous berry, 1–2-seeded.

Rare in Veligonda hills of Nellore district.
Fl. & Fr.: October–January.

Sanskrit: Telugu:

Medicinal uses: The roots, seeds, bark and wood contain alkaloids brucine and strychnine. The plant is used in medicine for the same purposes as *Strychnos nux-vomica.*

Strychnos nux-vomica L.

Family: Loganiaceae.

Deciduous tree, up to15 m tall, often with short, strong axillary spines. Leaves elliptic-orbicular, coriaceous. Flowers greenish-white, fragrant, in terminal compound cymes. Fruit globose berry, thick shelled, orange when ripe, slightly rough, but shining, seeds many.

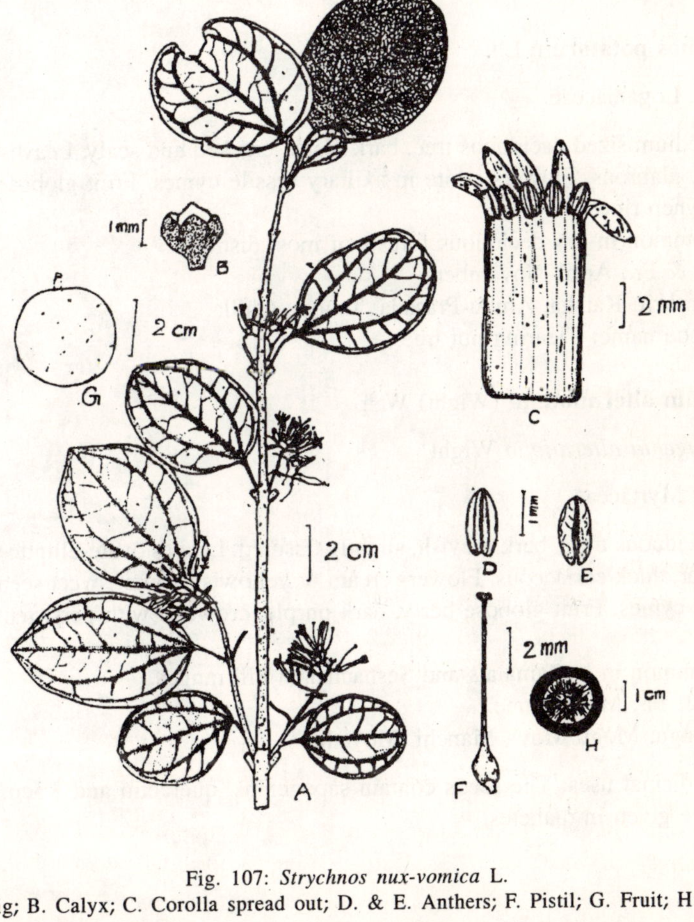

Fig. 107: *Strychnos nux-vomica* L.
A. Twig; B. Calyx; C. Corolla spread out; D. & E. Anthers; F. Pistil; G. Fruit; H. Seed.

Common in deciduous forests of most districts.
Fl. & Fr.: March–August.
Sanskrit: Karaskara, Chipita. Telugu: Mushti, Mushini.
Trade name: Snake wood, poison nut.

Medicinal uses: Seeds are used as a bitter stomachic and tonic. They stimulate respiratory and cardio vascular systems. Brucine has less physiological effects but four times more bitter than strychnine. Brucine is used in epilepsy. The seeds are used in rat-bites, venomous reptile bite, and rheumatism and alcoholism. Due to extreme toxicity strychnine is rarely used in medicine.

Active principle: Seeds contain 1.5 to 5% of bitter indole alkaloids. Chief constituents are strychnine and brucine. Vomicine a-and b-colubrine, Icajine, novacine, N Oxystrychnine, Pseudostrychnine and strychnicine are also present. Strychnine amounts to 30–50% of total alkaloids. It is concentrated towards the centre of the seed. Brucine is located towards the periphery.

Strychnos potatorum L.f.

Family: Loganiaceae.

Medium sized deciduous tree, bark black, cracked and scaly. Leaves ovate-elliptic, glabrous. Flowers white in axillary sessile cymes. Fruit globose berry, black when ripe.
Common in dry deciduous forests of most districts.
Fl. & Fr.: April–November.
Sanskrit: Kataka, Ambu-Prasada. Telugu: Chilla.
Trade name: Clearing nut tree.

Syzygium alternifolium (Wight) Walp.

Syn: *Eugenia alternifolia* Wight

Family: Myrtaceae.

Deciduous trees, bark greyish, slightly fissured. Leaves ovate-elliptic to sub-orbicular, thick coriaceous. Flowers cream or yellowish white, sweet scented, in axillary cymes. Fruit globose berry, dark purple, crowned with persistent calyx tube.
Common in Nallamalais and Seshachalam hill ranges.
Fl & Fr.: March–June.
Telugu: Mogi. Movi, Manchi, Moyadi.

Medicinal uses: The seeds contain sapogenins, quercetin and Kaempferol. They are given in diabetes.

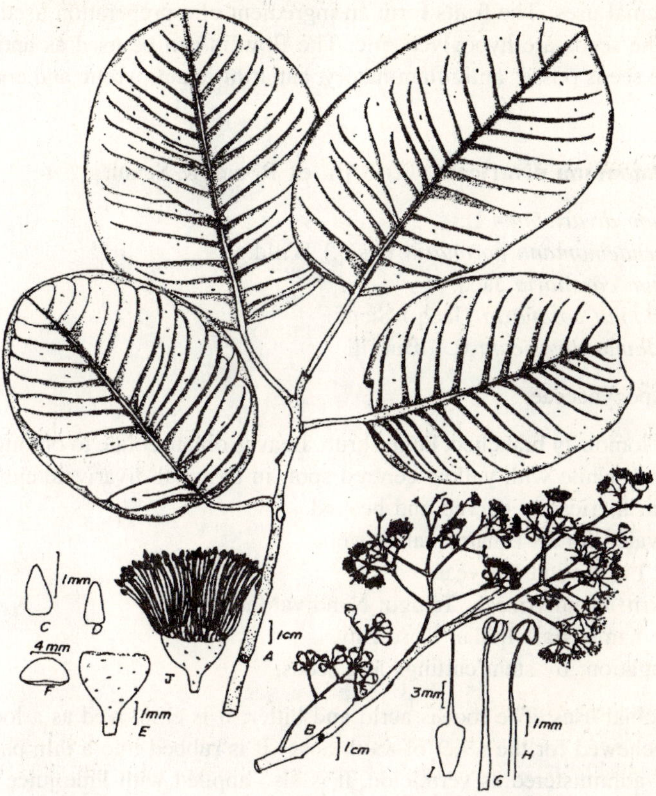

Fig. 108: *Syzygium alternifolium* (Wight) Walp.
A. Twig; B. Flowering branch; C. Bract; D. Bracteole; E. Calyx; F. Calyptra; G. & H. Stamens; I. Pistil; J. Flower.

Syzygium cumini (L.) Skeels

Syn: *Syzygium jambolanum* (Lam.) DC.
Myrtus cumini L.
Eugenia jambolana Lam.

Family: Myrtaceae.

Tall tree, bark ashy brown, smooth. Leaves opposite decussate, elliptic-oblong or lanceolate, coriaceous, glabrous. Flowers pale greenish white in axillary or terminal trichotomous panicles. Fruit berry, ovate or oblong, shining bright to dark when mature, 1-seeded.

Common in all forests, planted by forest department.

Fl. & Fr.: April–June.

Sanskrit: Jambuh. Telugu: Neredu.

Trade name: Jaman, Black Plum.

Medicinal uses: The fruits form an ingredient of a preperation used against asthma. The seeds are hypoglycaemic. The flowers can be used as antifertility agent. The seeds posses antiinflammatory, antiarthrtic antipyretic and anaelgesic activity.

Tabernaemontana divaricata (L.) R.Br. ex Roem. & Schult.

Syn: *Nerium divaricatum* L.
Tabernaemontana coronaria (Jacq.) Willd.
Nerium coronaria Jacq.
Ervatamia coronaria (Jacq.) Stapf
Ervatamia divaricata (L.) Burkill

Family: Apocynaceae.

Dichotomously branched, large shrub. Leaves oblanceolate to oblong. Flowers creamish-white with yellow centred spot, in terminal divarivate cymes, fragrant. Fruit mericarps, curved and beaked.

Cultivated for its ornamental flowers.

Fl. & Fr.: Round the year.

Sanskrit: Nandivriksah. Telugu: Nandivardhanamu.

Trade name: East Indian Rose Bay.

Propagation: by stem cuttings and seeds.

Medicinal uses: The root is acrid and bitter; it is employed as a local anodyne and chewed for the relief of tooth-ache. It is rubbed into a thin paste with water and administered as vermicide. It is also applied with lime juice to clear opacity of the carnea. The root charcoal and the milky juice of the leaves are used in opthalmia. The juice of flowers mixed with oil is employed for relieving the burning sensation in sore eyes, it is applied also in skin diseases. The milky juice of the leaves is used as a cooling application for wounds to prevent inflammation. The wood is used as a refrigerant.

Tacca leontopetaloides (L.) O.Kuntze

Syn: *Tacca pinnatifida* Forst.
Leonitice leontopetaloides L.

Family: Taccaceae.

A herb with globose root stock. Leaves in out line 30–90 cm in diameter circular, 3-partite, the segments variously pinnatifid. Flowers green tinged with purple. Fruit globose or ovoid berry.

Common in forests.

Fl. & Fr.: June–November.

Medicinal uses: The tubers are used in the treatment of piles. A bitter extract prepared by washing the grated tubers in running water is a rubifacient and is also given in diarrhoea and dysentery.

Tamarindus indica L.

Family: Caesalpiniaceae.

Trees, about 15–30 m tall, bark fissured. Leaves even pinnate, leaflets 12–16 pairs, oblong. Flowers yellow with pink striations in terminal few-flowered racemes. Pod linear-oblong, falcate, with brittle epicarp, pulpy mesocarp and leathery endocarp, seeds 1-many, quadrangular, brown, shining.

Common both in forests and plains, also planted in topes and avenue tree.

Fl. & Fr.: March–September.

Sanskrit: Chincha, Tintrini. Telugu: Chintachettu.

Trade name: Tamarind tree.

Propagation: by seeds.

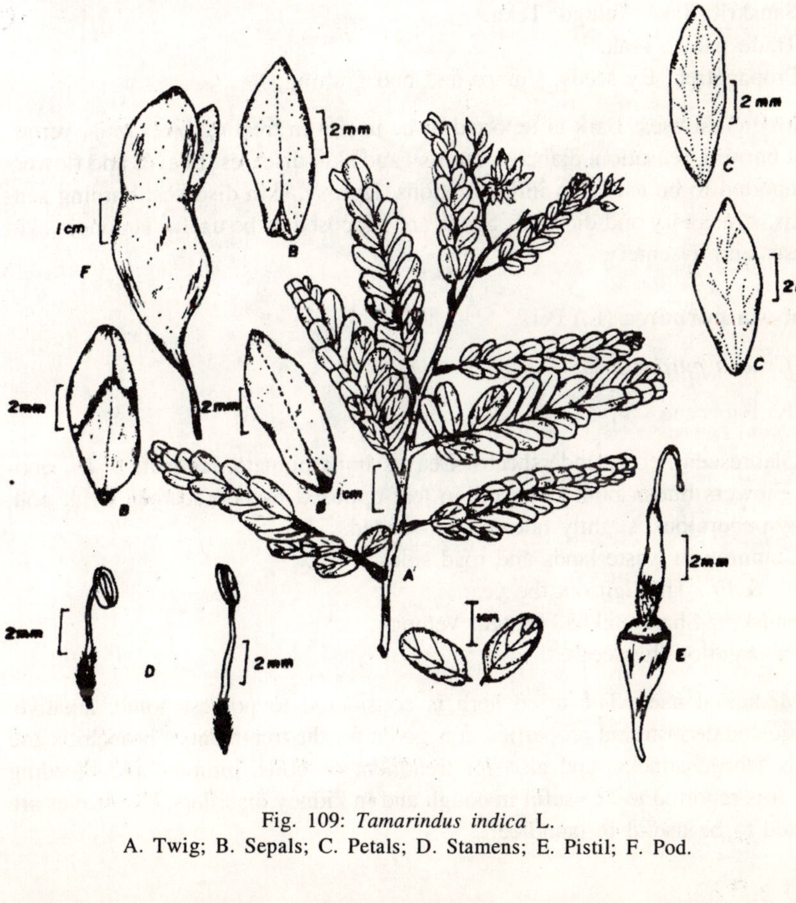

Fig. 109: *Tamarindus indica* L.
A. Twig; B. Sepals; C. Petals; D. Stamens; E. Pistil; F. Pod.

Medicinal uses: The root bark is used against asthma, amenorrhoea, gingivitis and ulcers. Leaves are useful for ulcers, fever, hilminthiasis, wounds, jaundice scabies, tumors, ringworms, boils, small pox and conjunctivitis. Fruits are laxative and also used opthalmopathy. Seeds are used to treat dysentery, burning sensation, giddiness, inflammations, chronic ulcers, abscess, haemorrhoids, vaginopathy, diabetes and general debility.

Tectona grandis Lf.

Family:Verbanaceae

Large deciduous trees, branchlets quadrangular, stellate tomentose. Leaves very large, broadly elliptic, glabrous above, stellate pubescent below. Flowers yellowish in terminal dichotomous, large panicles. Fruit drupe, enclosed in enlarged calyx.

Common through out the state in forests.
Fl. & Fr.: June–January
Sanskrit: Saka. Telugu: Teku.
Trade name: Teak.
Propagation: By seeds, stumps and bud grafting.

Medicinal uses: Bark is reported to be useful in treating bronchitis, verminosis, burning sensation, diabetes, leprosy and skin diseases. Leaves and flowers are reported to be useful in inflammations, leprosy, skin diseases, burning sensations, stranguary and diabetes. Seeds are supposed to be useful in cough, skin diseases and dysentery.

Tephrosia purpurea (L.) Pers.

Syn: *Cracca purpurea* L.

Family: Fabaceae

Glabrescent, erect undershrubs. Leaves imparpinnate, leafleats 9–19, obovate. Flowers bluish pink to purple in leaf-opposed pseudoracemes. Fruit pod, downy-puberulous, slightly falcate, 5–7-seeded.

Common in waste lands and road sides.
Fl. & Fr.: Through out the year.
Sanskrit: Sharapunkha. Telugu: Vempali.
Propagation: by seeds.

Medicinal uses: The dried herb is considered to possess tonic, laxative, diuretic and deobstruent properties. It is given for the treatment of bronchitis and bilious febrile attacks, and also for treatment of boils, pimples and bleeding piles. It is reported to be useful in cough and in kidney disorders. The leaves are reported to be useful in jaundice.

Fig. 110: *Tephrosia purpurea* (L.) Pers.
A. Twig.

Tephrosia villosa (L.) Pers.

Syn: *Galega villosa* L.
Tephrosia hitra (Buch.-Ham.) Benth.
Galega hitra Buch.-Ham.

Family: Fabaceae (Papilionaceae).

Erect, silky-villous undershrubs. Leaves imparipinnate, leaflets 5–7 pairs, obovate-oblanceolate. Flowers pink in axillary racemes. Fruit pod, seeds 6–8, strophiolate.

Common in waste lands and road sides.

Fl. & Fr.: June–October.

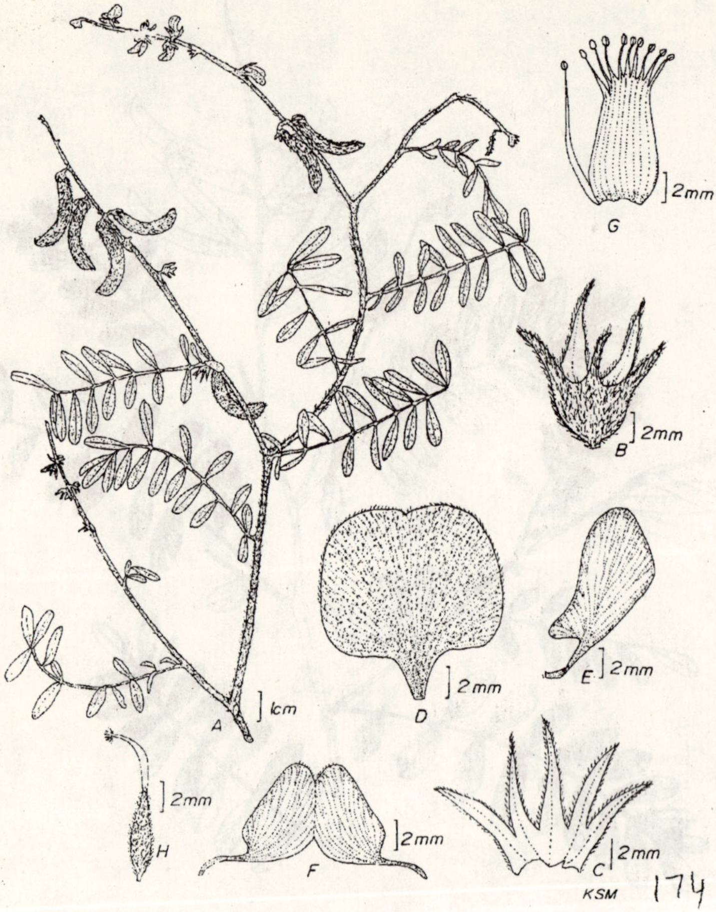

Fig. 111: *Tephrosia villosa* (L.) Pers.
A. Twig; B. Calyx; C. Calyx split open; D. Standard petal; E. Wing petal.

Medicinal uses: The juice of the leaves is given in dropsy. The fresh root is credited with hypoglycaemic properties. The leaves are used to treat dropsy and diabetes.

Terminalia arjuna (Roxb. ex DC.) Wight. & Arn.

Syn*:* *Pentaptera arjuna* Roxb. ex DC.

Family: Combretaceae.

Large deciduous tree, bark white and smooth exuding red resin which turns black when exposed. Leaves large, oblong, glabrous. Flowers dull yellow, in axillary spikes. Fruit obovoid-oblong drupe, 5-angled, 5-winged, woody.

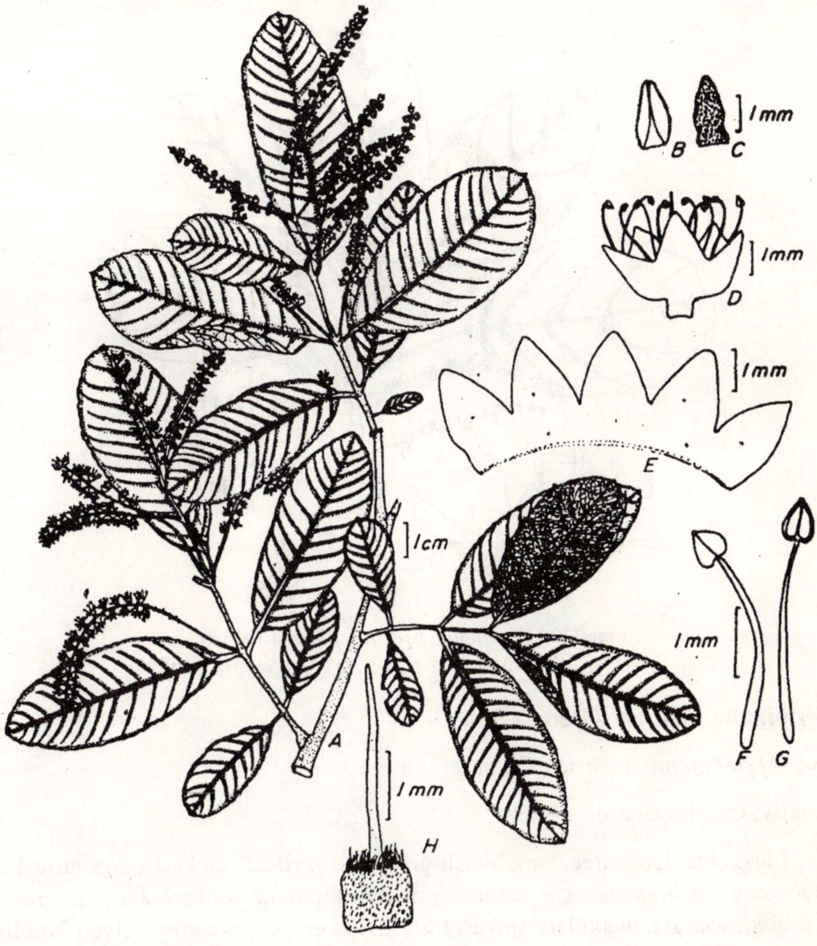

Fig. 112: *Terminalia arjuna* (DC.) Wight & Arn.
A. Twig; B. & C. Bracteoles; D. Flower; E. Calyx split open; F. & G. Stamens; H. Pistil.

Common all over the state along streams, river banks and water courses.

Fl. & Fr.: April–December.

Sanskrit: Arjuna. Telugu: Tella maddi.

Trade name: Arjun.

Propagation: by seeds.

Medicinal uses: Bark is used for fractures, ulcers, discharge of urethra, leucorrhoea, diabetes, anemia, cardiopathy, hyperhydrosis; asthma, bronchitis, states of exhaustion, tumors, dystentery, internal and external hemorrhage, liver cirrhosis and high blood pressure.

Active principe: Tannins, Steroids, Triterpenes and flavonoids.

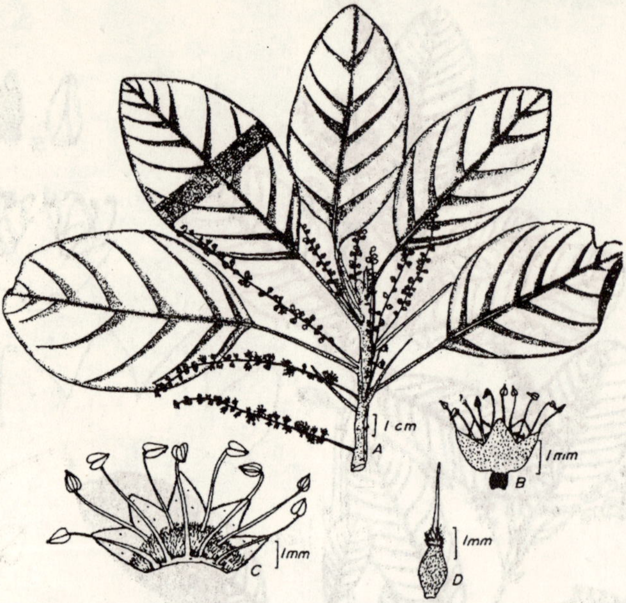

Fig. 113: *Terminalia bellirica* (Gaertn.) Roxb.
A. Twig; B. Flower; C. Calyx with stamens; D. Pistil.

Terminalia bellirca (Gaertn.) Roxb.

Syn: *Myrobalanus bellirica* Gaertn.

Family: Combretaceae.

Large deciduous tree, bark bluish-grey with vertical cracks. Leaves crowded at the ends of branches, coriaceous, broadly elliptic or obovate. Flowers cream coloured, scented, in axillary spikes. Fruit ovoid drupe, with grey-velvety woolly hairs.

Common in deciduous forests all over the state.

Fl. & Fr.: April–December.

Sanskrit: Vibhitaka. Telugu: Tani, Thandra.

Trade name: Beleric Myrobalan.

Propagation: by seeds.

Medicinal uses: The fruits are useful in stomach disorders such as indigestion, diarrhoea. It is given also as a brain tonic and is applied on eyes as soothing lotion. It is also useful in piles, leprosy, dropsy and fever. The half-ripe fruit is considered to be purgative but the ripe fruit has the opposite property.

The beleric myrobalan fruit is one of the three constituents of the famous Indian preperation Triphala, the other two being *Phyllanthus emblica* and *Terminalia chebula.*

Terminalia catappa L.

Family: Combretaceae

Deciduous trees, up to 1.5 m tall. Leaves densely clustered at the ends of branchlets, alternate, obovate. Flowers cream-coloured in axillary spikes. Fruit ellipsoid, 2-ridged.

Planted in gardens and as an avenue tree.

Fl. & Fr.: September–June.

Sanskrit: Grahadruma. Telugu: Badam.

Trade name: Indian Almond Tree.

Propagation: by seeds.

Medicinal uses: The leaves act as a sudorific, and they are applied to rheumatic joints. The juice of the young leaves is used to prepare an ointment for scabies, leprosy and other cutaneous diseases. The bark is endowed with mild diuretic and fairly potent cardiotonic properties. It contains tannin, which is used as an astringent in dysentery.

Terminalia chebula Retz.

Syn: *Myrobalanus chebula* (Retz.) Gaertn.

Family: Combretaceae.

A moderate sized deciduous tree, up to 20 m tall. Leaves subopposite, ovate, with two glands near base. Flowers dull white in simple spikes, fragrant. Fruit ovoid drupe, yellowish-green.

Occasional in deciduous forests.

Fl. & Fr.: March–December.

Sanskrit: Haritaki, Harra, Harada, Abhaya. Telugu: Karakkaya.

Trade name: Chebulic Myrobalan, Ink tree.

Propagation: by seeds.

Medicinal uses: The dried fruits paste is applied on chronic ulcers, wounds and sealds. It is used as a gargle in inflammation of mucous membrane of mouth. Fruits are also used as laxative. They have been found to have some effect on blood pressure as cardiac tonics. The powder of the fruit is used as a dentifries for the strength of gums.

Terminalia pallida Brandis

Family: Combretaceae.

Small ever green trees. Leaves small, coriaceous, clustered at the ends of branches, ovate. Flowers in simple spikes. Fruit ovoid drupe, very faintly 5-ridged.

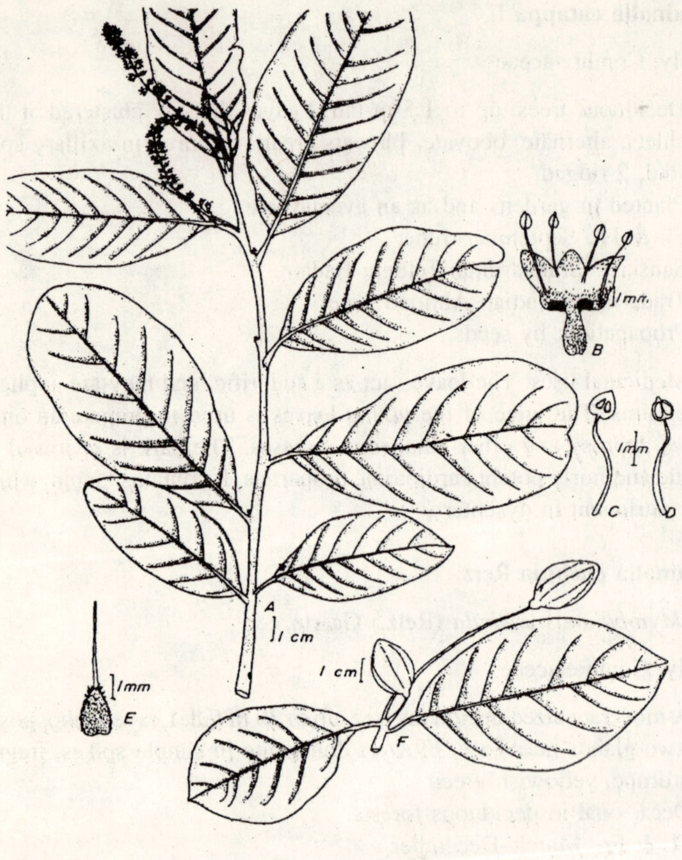

Fig. 114: *Terminalia chebula* Retz.
A. Twig; B. Part of Calyx with stamens; C. & D. Stamens; E. Pistil; F. Fruiting branch.

In dry deciduous forests of Cuddapah and Chittor districts.
Telugu: Tella karaka.
Trade name: White Gall Nut.

Medicinal uses: Bark is used as diuretic.

Thespesia populnea (L.) Soland. ex Corr.

Syn: *Hibiscus populneus* L.

Family: Malvaceae.

Trees, up to 10 m tall, branchlets clothed with peltate scales. Leaves ovate, entire, stellate tomentose beneath. Flowers solitary, axillary, yellow. Fruit indehiscent capsule.

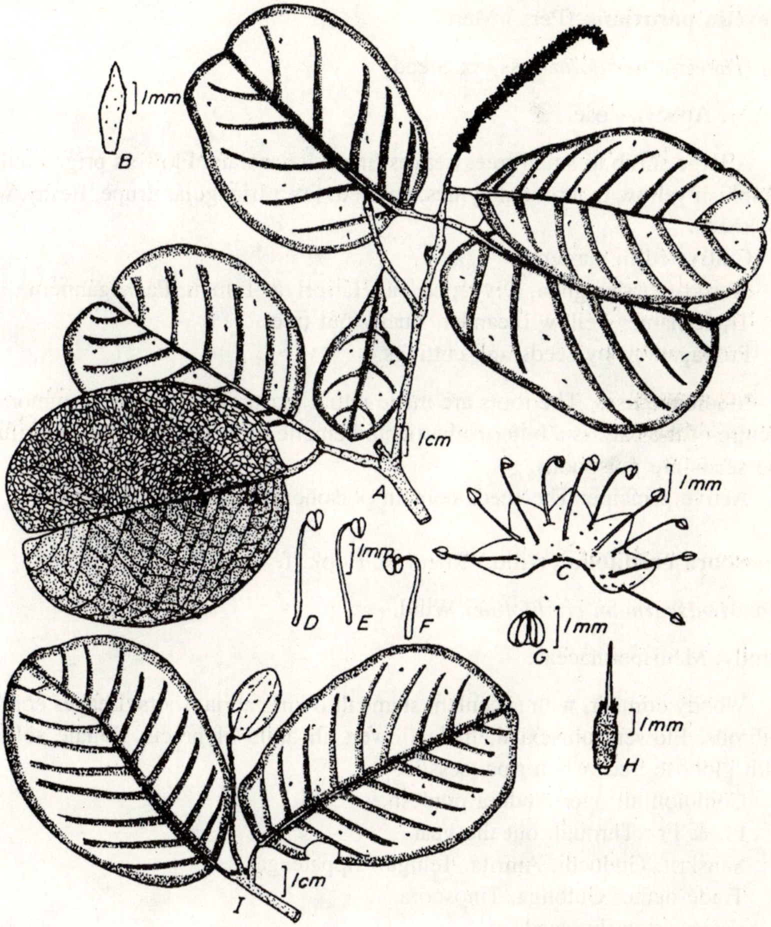

Fig. 115: *Terminalia pallida* Brandis
A. Twig; B. Bract; C. Calyx with Stamens; D., E. & F. Stamens; G. Anther; H. Pistil; I. Fruiting twig.

Often planted in gardens and on road sides.
Fl. & Fr.: Through out the year.
Sanskrit: Harpiccha. Telugu: Gangaravi, Gangarenu.
Trade name: Portia Tree. Indian Tulip Tree.
Propagation: by seeds.

Medicinal uses: Different parts have been used to treat various skin ailments like scabies, psoriasis, wounds and ulcers. The astringent bark, roots and fruits are stated to be used in dysentey, cholera and haemorrhoids. A compound oil of bark and capsule is useful in urethritis and gonorrhoea.

Thevetia peruviana (Pers.) Merr.

Syn: *Thevetia neriifolia* Juss. ex Steud.

Family: Apocynaceae.

A large shrub to small tree. Leaves linear-lanceolate. Flowers bright yellow or pinkish yellow in terminal cymes, scented. Fruit triangular drupe, fleshy, with 2–4 seeds.

Cultivated in gardens.

Sanskrit: Ashvaghna, Divyapushpa, Haripriya. Telugu: Pachaganneru.

Trade name: Yellow oleander, Lucky nut tree.

Propagation: by seeds and cuttings.

Medicinal uses: The roots are made into a plaster and applied to tumors. A tincture of the bark is a bitter cathartic and emetic, it is also used as a febrifuge. The seeds are poisonous.

Active principe: The seeds contain poisonous cardiac gycosides.

Tinospora cordifolia (Willd.) Miers ex Hook. f.

Syn: *Menispermum cordifolium* Willd.

Family: Menispermaceae.

Woody climber, with succulent stems throwing aerial roots. Leaves cordate, glabrous. Flowers unisexual, male flowers in axils of bracts, female solitary. Fruit globose, red, when ripe, sessile.

Common all over Andhra pradesh.

Fl. & Fr.: Through out the year.

Sanskrit: Guduchi, Amrita. Telugu: Tippateega.

Trade name: Gulonga, Tinospora.

Propagation: by seeds.

Medicinal uses: Whole plant possess antiallergic, antidiabetic, antihepatotoxic, antipyretic, antispasmodic, antiinflammatory and anticancer properties. It is bitter in taste and considered tonic, stomachic, anaelgesic, antiperodic and diuretic. It is also used in general debility, dyspepsia, fevers and urinary diseases. It is also considered as aphrodisiac. The dry stem, with bark intact, constitute the drug.

Active principle: Stem contains glucoside, alkaloids Berberine, tinosporine, giloinin and gilonin.

Toddalia asiatica (L.) Lam.

Syn: *Toddalia aculeata* Pers.

Family: Rutaceae.

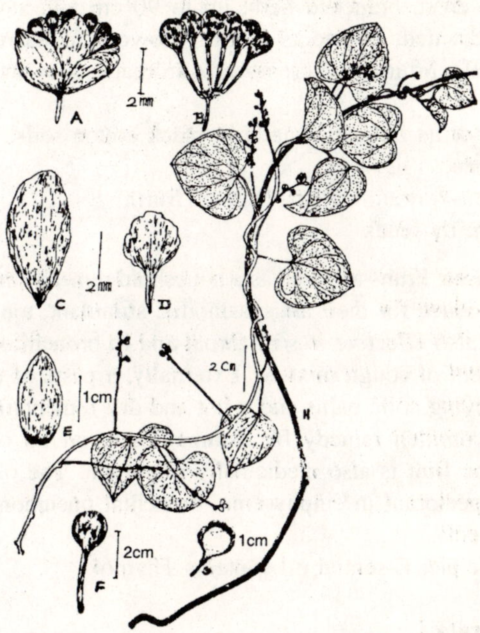

Fig. 116: *Tinospora cordifolia* (Willd.) Hook. f. & Thoms.
A. Flower; B. Androecium; C. Sepal; D. Petal; E. Bract; F. Stamen; G. Fruit; H. Twig.

Armed straggler, branchlets glandular-tomentose, terminal prickles recurved. Leaves 3-foliolate, sessile. Flowers in axillary panicles, unisexual. Fruit berry, 1–5-seeded.

Common in forests in most districts.

Fl. & Fr.: July–February.

Sanskrit: Tikshanakshi, Kanchan. Telugu: Kondakasinda, Mirapakandra.

Trade name: Forest pepper, Lopez Tree, Wild orange Tree.

Propagation: by seeds.

Medicinal uses: The root bark is credited with diaphoretic, stomachic, antipyretic and antimalarial properties.

Active principle: Root bark contains alkaloids cheberythrine (toddaline), toddalinine etc.

Trachyspermum ammi (L.) Sprague ex Turril.

Syn: *Carum copticum* Hiern.
Trachyspermum copticum Link

Family: Apiaceae (Umbelliferae)

An annual, erect, branched herb, up to 90 cm tall, stems striate. Leaves 2–3-pinnately divided, segments linear. Flowers in terminal, pedunculate, compound umbels. Fruit ovoid cremocarp, aromatic, greyish brown, surface tubercular.

Cultivated for its fruit, especially in black cotton soils.

Telugu: Vamu.

Trade name: Ajowan, Carum, Omum (fruit).

Propagation: by seeds.

Medicinal uses: Fruits are used as a household remedy for indigestion. Fruits are also much valued for their antispasmodic, stimulant, tonic and carminative properties. It is also effective in sore throat and in bronchitis, and often constitutes an ingredient of cough mixture. Externally, a paste of the crushed fruit is applied for relieving colic pains and a hot and dry fomentation of the fruits on the chest is a common remedy for asthma. Essential oil obtained by steam-distillation of the fruit is also medicinally important. The oil is reported to be useful as an expectorant in emphysema, bronchial pneumonia and some other respiratory ailments.

Active principle: Essential oil contains Thymol.

Tragia involucrata L.

Family: Euphorbiaceae

Climbing or twining hispid herbs with stinging hairs. Leaves linear-lanceolate or linear-elliptic. Flowers in axillary and terminal racemes, unisexnal, male flowers many from middle above, female flowers few, near the base. Fruit capsule, hispid.

Rare in forests in coastal and Rayalaseema districts.

Fl. & Fr.: December–March

Sanskrit: Dhusparsha, Vrischikali. Telugu: Theegaduradagunta, Chinnaduradagunta.

Trade name: Indian stinging nettle.

Propagation: by seeds.

Medicinal uses: The roots possess diaphoretic and alterative properties and are given in fever when the extremities are cold and also for pains in the legs and arms. A paste prepared from the roots is applied to aid the extraction of guineaworms. The root is also used in old venereal complaints and also as a blood purifier. The fruit is a constituent of a medicinal preperation, known as ksharacanda, used in cases of enlarged spleen. The fruits are rubbed over the head with water to cure baldness.

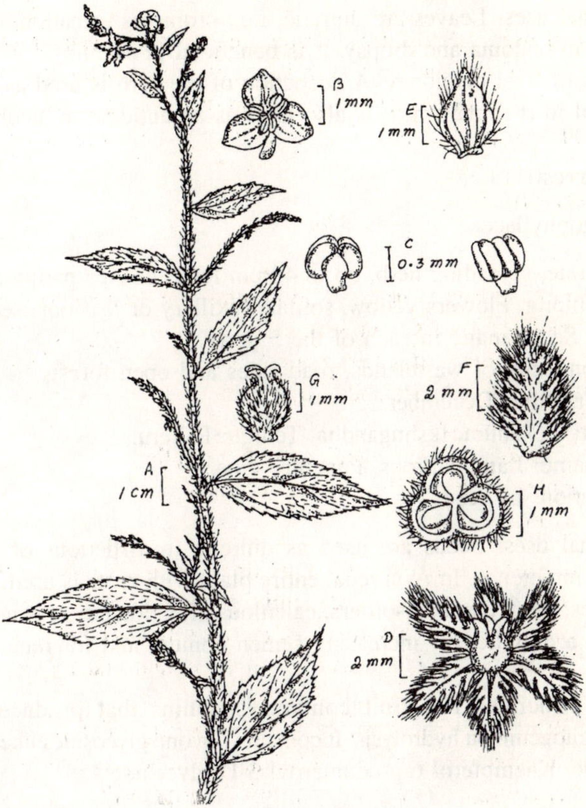

Fig. 117: *Tragia involucrata* L.
A. Twig; Male: B. Flower; C. Anthers; D. Flower; E. Bract; F. Involucre lobe; G. Pistil; H. Ovary T.S.

Trianthema portulacastrum L.

Syn: *Trianthema monogyna* l.

Family: Aizoaceae

An annual prostrate, succulent herb. Leaves opposite, very unequal, leaf blade obovate-orbicular or oblong. Flowers bright pink or white, axillary, solitary. Fruit capsule, circumscissile, seeds 2, dull black.

Commed weed of wastelands and road sides.

Fl. & Fr.: July–November.

Sanskrit: Shvetapunarnava. Telugu: Galijeru.

Trade name: Horse purslane.

Propagation: by seeds.

Medicinal uses: Leaves are diuretic, i.e., promotes urination and is, therefore, useful in oedoma and dropsy. It is beneficial in swellings of body caused by disorders of liver or kidney. A decoction of the herb is used as a vermifuge and is useful in rheumatism, it is also used as an antidote to alcoholic poison.

Tribulus terrestris L.

Family: Zygophyllaceae

A prostrate, spreading herb, up to 40 cm long. Leaves paripinnate, leaflets 4–7 pairs, oblong. Flowers yellow, solitary, axillary or leaf-opposed. Fruit globose, spiny. Seeds many in each of the 5 parts.

Common weed of wastelands, road sides and open forests.

Fl. & Fr.: June–December.

Sanskrit: Gokshura, Ikshugandha. Telugu: Palleru.

Trade name: Land caltrops, Puncture vine.

Propagation: by seeds.

Medicinal uses: Seeds are used as diuretic in afflictions of urinary discharges and impotency. In Ayurveda, entire plant with roots is used. It is used in kidney stones, gout, uterus disorders, calculous affection etc. It is used in Unani medicine as a diuretic, for increasing semen content and for removing kidney stones.

Active principle: The fruit contains saponins that produce diosgenin, ruscogenin, gitogenin on hydrolysis. It contains flavone glycosides like Kaempferol 3-rhamnoside, Kaempferol 6-p-coumaroyl 3-D glycoside.

Trichodesma indicum (L.) R.Br.

Syn: *Borago indica* L.

Family: Borginaceae.

An annual, erect, hispid herb. Leaves sessile, ovate-oblong. Flowers pinkish-violet, solitary, axillary. Fruits pyramiadal, 4-ribbed, nutlets ovoid.

A weed in waste places.

Fl. & Fr.: July–March.

Sanskrit: Adhapushpi. Telugu: Guvvagutti.

Propagation: by seeds.

Medicinal uses: The herb is credited with emollient and diuretic properties and is used for making emollient poultices. It is prescribed by the Ayurvedic physicians for the expulsion of dead foetus. An infusion of the leaves is considered depurative. The root is pounded and made into a paste for application on the swellings particularly of the joints. The root is also used for the treatment of dysentery and fever. The flowers are reported to be employed as sudorific and pectoral.

Trichosanthes cucumeriana L.

Family: Cucurbitaceae.

Annual climber, branches sparsely puberulous, tendrils slender, branched. Leaves orbicular-reniform or broadly ovate, angular or 5–7-pamately lobed, densely glandular pubescent below. Flowers white, male flowers in racemes, female flowers solitary. Fruit berry, glabrous, green and striped when fresh, scarlet or orange when ripe, seeds many, flattened.

Occasional throught out the state.

Sanskrit: Amritaphala, Kashtbanjan. Telugu: Chetipotla.

Propagation: by seeds

Medicinal uses: Both the root and fruit are considered cathartic. The root is used as a cure for bronchitis, headache and boils. The leaves are used in biliousness, their juice is emetic. The seeds are considered antifebrile and anthelmintic.

Tridax procumbens L.

Family: Asteraceae (Compositae)

Annual creeping and ascending, hispid herbs. Leaves ovate, strigose on both sides. Flowers yellow in terminal heads. Fruit obconical achene, ribbed.

Common weed of wastelands and road sides.

Fl. & Fr.: Through out the year.

Telugu: Gaddichamanthi.

Trade name: Mexican Daisy,

Propagation: by seeds.

Medicinal uses: The leaves are reported to be employed in bronchial catarrh, dysentery and diarrhoea. The leaf juice possesses antiseptic properties, it is used to check haemorrhage from cuts, bruises and wounds.

Trigonella foenum-graecum L.

Family: Fabaceae (Papilionaceae).

An annual aromatic herb. Leaves pinnately 3-foliolate, leaflets oblanceolate-oblong. Flowers white or creamish, axillary, solitary. Fruit pod, 10–20-seeded, seeds greenish brown, oblong, with a deep groove across on corner.

Cultivated for its seeds and also as leafy vegetable.

Sanskrit: Methika, Chandrika, asumodhagam. Telugu: Menthulu (seeds), Menthiaku (leaves).

Trade name: Fenugreek.

Propagation: by seeds.

Medicinal uses: The seeds are aromatic, carminative, tonic and galactogogue. They are used externally in poultice for boils, abscesses and ulcers and internally as emollient for inflammations of the intestinal tract.

Triumfetta rhomboidea Jacq.

Syn: *Triumfetta angulata* Lam.

Family: Tiliaceae.

A perennial much branched, erect pubescent under shrubs, up to 2 m tall. Leaves polymorphous, lower leaves palmately 3-lobed, upper ones rhomboid to narrow lanceolate or elliptic, stellate pubescent. Flowers yellow, in dense cymes. Fruit globose capsule, clothed with smooth hooked spines, pea sized.

Common weed of waste places and hedges.

Fl. & Fr.: August–January.

Sanskrit: Jhinjharita. Telugu: Chiru Sitrika.

Trade name: Bur Bush, Bur weed.

Propagation: by seeds.

Medicinal uses: The root is used in dysentery and the bark and fresh leaves in diarrhoea. Pounded roots are given for the treatment of intestinal ulcers and their hot infusion is said to hasten parturition. The leaves and the flowers are used against leprosy.

Tylophora fasciculata Buch.-Ham. ex Wight

Family: Asclepiadaceae

A slender, suffruticose, tuberous undershrub with watery sap. Leaves ovate-lanceolate, coriaceous, glabrous above, pubescent on nerves beneath. Flowers small, chocolate brown, in lateral, peduncled, umbellate or racemose cymes. Fruit paired follicles, ovate or spindle-shaped, glabrous.

Coastal and Telangana districts in forests.

Fl. & Fr.: June–December.

Propagation: by seeds.

Medicinal uses: Roots and leaves possess emetic and purgative properties. They are reported to contain an amorphous alkaloid and are used as a substitute for *Cephealis ipecacuanha*. Juice of the roots, mixed with milk is given as a tonic. A decoction of the roots in given as a tonic. A decoction of the roots in given in fevers.

Tylophora indica (Burm.f.) Merr.

Syn: *Tylophora asthmatica* (L.f.) Wight & Arn.
Cynanchum indicum Burm.f.
Asclepias asthmatica L.f.

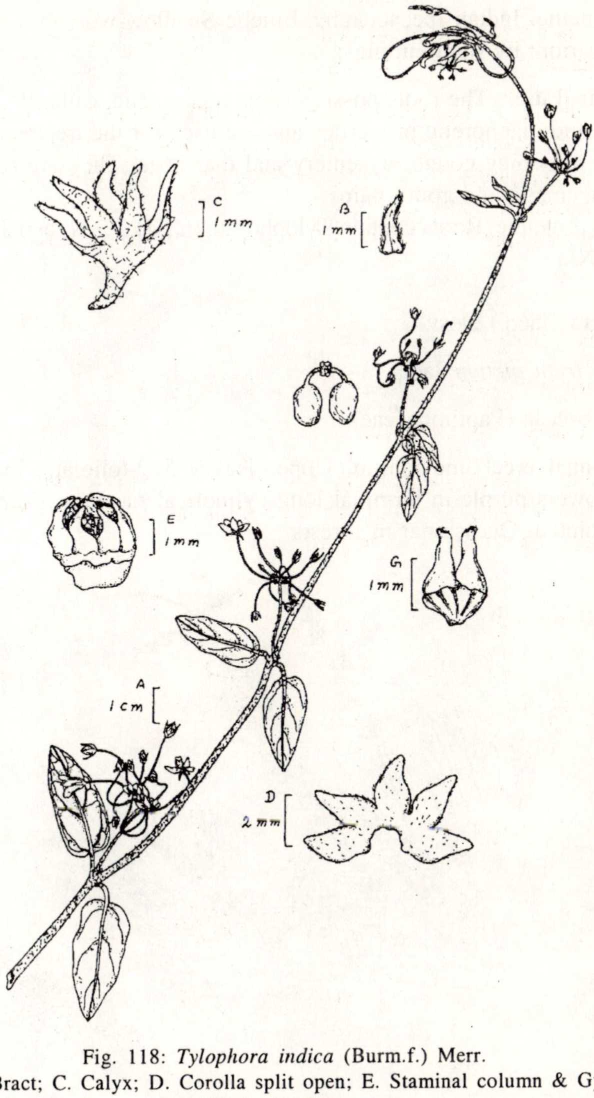

Fig. 118: *Tylophora indica* (Burm.f.) Merr.
A. Twig; B. Bract; C. Calyx; D. Corolla split open; E. Staminal column & Gynostegium; F. Pollinia; G. Gynoecium.

Family: Asclepiadaceae.

Much branched climber with long fleshy roots and watery latex. Leaves cordiform, oblong to lanceolate, opposite. Flowers pale yellow, purple within, in lateral cymes. Fruits in pairs, ridged.

Common among bushes, in hedges and in open forest all over Andhra pradesh.

Fl. & Fr.: February–October

Sanskrit: Antamul. Telugu: Verripala, Kukkapala.

Trade name: Indian Ipecacuanha, Emetic Swallow-wort.

Propagation: by stem cuttings.

Medicinal uses: The roots possess stimulant, emetic, cathartic, expectorant, stomachic, and diaphoretic properties and are used for the treatment of asthma, bronchitis, whooping cough, dysentery and diarrhoea. They are reported to be given in rheumatic and gouty pains.

Active principle: Roots contain Tylophorine ($C_{24}H_{27}O_4N$) and Tylophorinine ($C_{23}H_{25}O_4N$).

Uraria picta (Jacq.) Desv.

Syn: *Hedysarum pictum* Jacq.

Family: Fabaceae (Papilionaceae)

An annual erect undershrub. Upper leaves 5–9-foliolate, leaflets linear-oblong. Flowers purple in terminal long cylindrical racemes. Fruit indehiscent pod, 3–6-jointed. Occasional in forests.

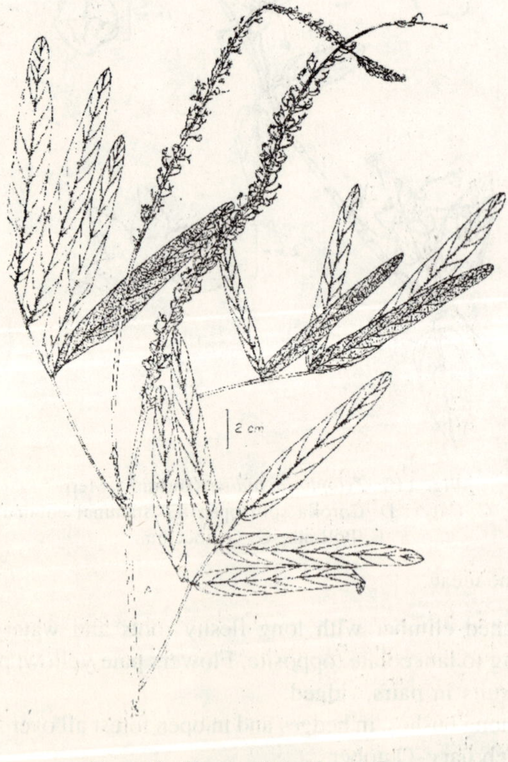

Fig. 119: *Uraria picta* (Jacq.) Desv.
A. Twig.

Fl. & Fr.: August–November.
Sanskrit: Prasniparni. Telugu: Pingerragadda.
Propagation: by seeds.

Medicinal uses: It is one of the components of Dasamoola. It is credited with fracture healing properties. Its total extract has been found to effect better and quicker healing of fractures in experimental animals due to early accumulation of phosphorus and more deposition of calcium.

Vanda tessellata (Roxb.) Hook.ex G. Don

Syn: *Epidendrum tessellatum* Roxb.
Vanda roxburgii R.Br.

Family: Orchidaceae.

A stout epiphyte, 30–60 cm long. Leaves linear, thick coriaceous, distichous, strap-shaped. Flowers grey, buf for red in axillary racemes. Fruit ribbed capsule.

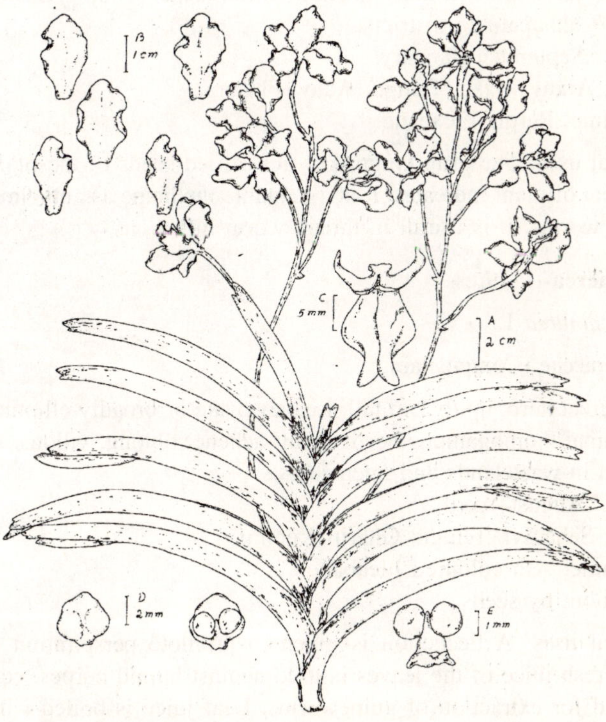

Fig. 120: *Vanda tessellata* (Roxb.) Hook. ex Don
A. Twig; B. Sepal; C. Petal; D. Lip front view; E. Operculum with pollinia; F. Pollinium.

Common epiphyte in forests.
Fl.: July–August, Fr.: August onwards.
Sanskrit: Rasna, Athirasa. Telugu: Kanapa, Chettubadanika.
Trade name: Ichneumon Plant.
Propagation: by seeds.

Medicinal uses: The leaves are pounded and the paste is applied to the body to bring down fever, their juice is dropped in the ear for the treatment of otitis and other inflammatory conditions. The roots are used in dyspepsia, bronchitis, rheumatism, fever and nervous troubles.

Vernonia anthelmintica (L.) Willd.

Syn: *Conyza anthelmintica* (L.) O.Kuntze
Centrantherum anthelminticum (Willd.) O.Kuntze

Family: Asteraceae (Compositae).

Erect, branched, hispid-pubescent herb to subshrub. Leaves elliptic. Flowers in heads, heads solitary, terminal, homogamous, pink. Fruit oblong achene, ribbed.

Rare in forests and waste lands in Kurnool, Cuddapah, Chittor, East Godavari, Medak and Visakhapatnam districts.
Fl. & Fr.: September–January.
Sanskrit: Aranyajiraka. Telugu: Adavi jilakara.
Trade name: Purple Fleabane.

Medicinal uses: The drug comprises fresh dried seeds of the plant. As the scientific name of plant indicates, it is a valuable medicine as anthelmintic, that is to destroy worms. It is useful in thread worm infections.

Vernonia cinerea (L.) Less.

Syn: *Conyza cinerea* L.

Family: Asteraceae (Compositae)

Annual erect herb, up to 1 m tall. Leaves ovate or broadly elliptic. Flowers pink, in terminal, homogamous heads. Fruit achene, oblong, villous.

Common in wastelands and open forests.
Fl. & Fr.: August–April.
Sanskrit: Sahadivi. Telugu: Ghariti Kamini.
Trade name: Ash coloured Fleabane.
Propagation: by seeds.

Medicinal uses: A decoction is given to promote perspiration in febrile conditions. Fresh juice of the leaves is used against humid herpes, eczema and ringworm and for extraction of guineworms. Leaf juice is boiled with oil and used for the treatment of elephantiasis. The seeds are commonly used as an anthelmintic and alexipharmic; they are said to be quite effective against round worms and thread worms.

Vetiveria zizanioides (L.) Nash

Family: Poaceae (Gramineae)

A perennial, tufted, robust grass, up to 2 m tall. Leafsheaths 5–20 cm long, glabrous, blades linear. Inflorescence panicle, spikelets lanceolate binate. Caryopsis linear-onlong.

Occasional grass in marshy areas.

Fl. & Fr.: August–March.

Sanskrit: Vetiver. Telugu: Vattivellu.

Trade name: Khus-Khus, Vetiver.

Propagation: by slips.

Medicinal uses: Oil is obtained by distillation of roots is known as vetiver oil. It is one of the most valuable and important raw material in perfumary. It is widely used in perfumes and cosmetics and for scenting soaps.

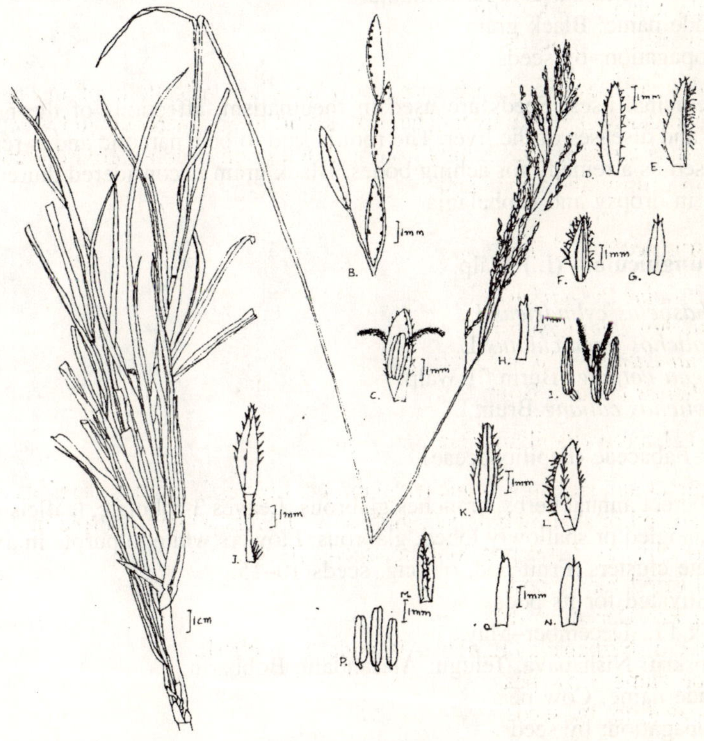

Fig. 121: *Vetiveria zizanioides* Nash.
A. Twig; B. Part of raceme; C. Sessile spikelet; D. Lower glume; E. Upper glume; F. lower lemma; G. & H. Upper lemma front and side views; I. Pistil with stamens; J. Pedicelled spikelet; K. Lower glume; L. Upper glume; M. Lower lemma; N. & O. Upper lemma front and side views; P. Stamens.

The oil is reported to be used as a carminative in flatulence, colic and vomting. It is regarded as a stimulant, diaphoretic and refrigerant. A decoction of leaves is applied topically in rheumatism, lumbago and sprain.

Vigna mungo (L.) Hepper

Syn: *Phaseolus mungo* L.
Phaseolus radiatus Roxb.

Family: Fabaceae (Papilionaceae)

Erect herbs, up to 30 cm, branches strigose. Leaves 3-foliolate, leaflets oblong-lanceolate or ovate. Flowers yellow in axillary or leaf-opposed in racemes. Fruit pod, hairy, about 10-seeded.

Cultivated for its seeds.
Fl. & Fr.: August–December.
Sanskrit: Masha. Telugu: Minumulu.
Trade name: Black gram
Propagation: by seeds.

Medicinal uses: Seeds are used in rheumatism, affections of the nervous system and diseases of the liver. The root is said to be a narcotic and is reported to be used as a remedy for aching bones. Black gram is considered diuretic and is used in dropsy and cephalagia.

Vigna unguiculata (L.) Walp.

Syn: *Phasoelus cylindricus* L.
Dolichos unguiculatus L.
Vigna catjang (Burm.f.) Walp.
Dolichos catjang Brum.f.

Family: Fabaceae (Papilionaceae)

Suberect annual herbs, branches glabrous. Leaves 3-foliolate, leaflets ovate-deltoid, angled or shallowly lobed, glabrous. Flowers white to purple in axillary umbellate clusters. Fruit pod, oblong, seeds 10–15.

Cultivated for its pods.
Fl.& Fr.: December–May.
Sanskrit: Nishapava. Telugu: Alasandalu, Bobbarlu.
Trade name: Cow pea.
Propagation: by seeds.

Medicinal uses: Eating a handful of roasted green seeds daily, ia a medicine for neuritis, insomnia, weakness of memory, indigestion, dyspepsia, sensation of pins and needles in the limbs, periodic palpitation, congestive cardiac failure etc. Decoction of the dry seeds, a glassful, drunk everyday cures hyperacidity, scanty urination, nausea and vomting due to presence of round worms etc. Eating boiled grains with cooked leaves of drumstick increases strength, improves digestion and acts as a very effective aphrodisiac. Infusion of the leaves is applied over the face to cure pimples.

Vitex negundo L.

Family: Verbenaceae.

Shrubs or small trees, tender parts grey pubescent, young branches 4-angled. Leaves 3–5-foliolate, leaflets oblong-oblanceolate, central leaflet larger. Flowers violet to purple in terminal or rarely axillary cymose panicles. Fruit globose drupe, black when ripe.

Common on the river banks or fencings.

Fl. & Fr.: May–February.

Sanskrit: Nirgundi. Telugu: Tellavavili.

Trade name: Five leaved chaste tree.

Propagation: by seeds and tender stem cuttings.

Medicinal uses: Leaves and roots constitute the drug. The leaves are applied to rheumatic swellings of joints, in sprains etc. The leaves are considered tonic and vermifuge. The leaves are smoked for the relief of headache and catarrh. The decoction of leaves and the vapours are employed in baths for the treatment of febrile, catarrhal and rheumatic affections. The roots possess tonic, febrifugal, expectorant and diuretic properties. They are used in dyspepsia and rheumatism. The powdered root is prescribed as an anthelmintic and as a demulcent.

Vitex trifolia L.

Family: Verbenaceae

A slender bushy shrub or small tree, tender parts softly pubescent. Leaves 1–3-foliolate, leaflets obovate-rhomboid. Flowers bluish in terminal panicles. Fruit drupe, purplish-black when ripe.

Occasional along streams.

Fl. & Fr.: Through out the year.

Sanskrit: Jala Nirgundi. Telugu: Niruvavili, Vavili.

Medicinal uses: The leaves are used in poultices for rheumatic pains, inflammation, sprains etc. The powdered leaves and their infusion are given in fever. A decoction of the roots is used as a febrifuge. The roots are reported to be useful in treating painful inflammations, cough and fevers. Flowers are effective in treating fevers and fruits in treating amenorrhoea.

Vitis vinifera L.

Family: Vitaceae.

A woody, climbing, tendril-bearing vine. Leaves orbicular, 3–5-lobed. Flowers in compound compact panicles.

Sanskrit: Draksha. Telugu: Draksha.

Trade name: Common Grape-Wine.

Propagation: by seeds.

Medicinal uses: Grapes, both fresh and dried, have varied uses in Ayurveda and Unani system of medicine. Fresh grapes are considered laxative, stomachic, diuretic, demulcent and cooling. Raisins (dried fruits) are also demulcent, laxative, cooling and expectorant. Grapes are used in the preperation of Chyavanaprasa. The juice of the unripe berries is used as an astringent in throat affections. The leaves are astringent and are some times used in diarrhoea. The sap of the young branches is reported ton be used for skin diseases and opthalmia.

Wattakaka volubilis (L.f.) Stapf

Syn: *Dregea volubilis* (L.f.) Benth. ex Hook.f.
Marsdenia volubilis (L.f.) T.Cooke

Family: Asclepiadaceae.

Extensive perennial, woody, pubescent twiner with watery latex. Leaves broad ovate to suborbicular. Flowers greeenish-yellow in drooping umbellate cymes. Fruit paired oblong follicles, seeds with white soft coma.

Common in hedges and open forests.

Sanskrit: Madhumalathi. Telugu: Dudhipala, Palateega.

Trade name: Sneezing silk cotton.

Propagation: by seeds, stem cuttings.

Medicinal uses: The leaves are employed in applications for boils and abscesses. The roots and tender stalks are considered emetic and expectorant.

Chemistry: The roots contain a glucoside and traces of an alkaloid.

Wedelia chinensis (Osbeck) Merr.

Syn: *Wedelia calendulacea* Less.
Solidago chinensis Osbeck.

Family: Asteraceae (Compositae).

Perennial procumbent or ascending herb, stems frequently rooting at nodes. Leaves oblanceolate. Flowers yellow, in terminal heterogamous heads. Fruit achene, triquetrous.

Rare in moist places.

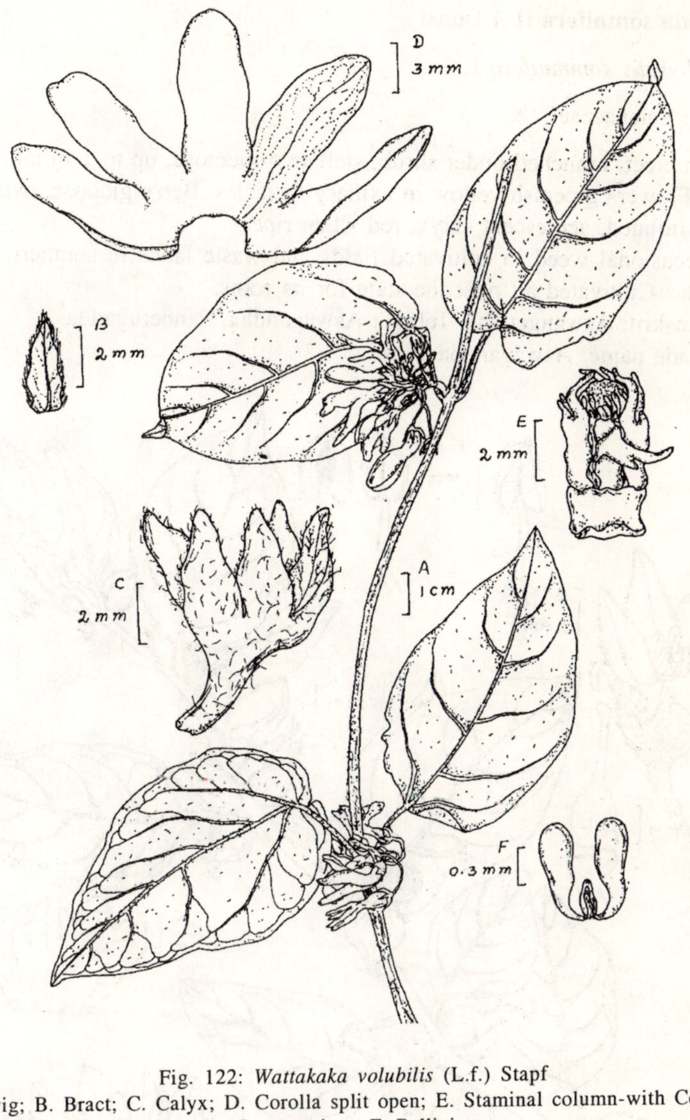

Fig. 122: *Wattakaka volubilis* (L.f.) Stapf
A. Twig; B. Bract; C. Calyx; D. Corolla split open; E. Staminal column-with Corona & Gynostegium; F. Pollinia.

Fl. & Fr.: June–December.
Sanskrit: Bhringaraja, Kesaraja, Pitabringi.
Propagation: by seeds.

Medicinal uses: The leaves are regarded as tonic and alterative, useful in cough, cephalagia and diseases of the skin. A decoction of the herb is used in uterine haemorrhage and menorrhagia.

Withania somnifera (L.) Dunal

Syn: *Physalis sommnifera* L.

Family: Solanaceae.

An erect, branched under shrub, stellate tomentose, up to 1 m tall. Leaves ovate. Flowers greenish yellow in axillary fascicles. Berry globose, overtopped by the inflated, accrescent calyx, red when ripe.

Occasional weed in cultivated fields and waste lands in southern Andhra pradesh. Cultivated all over the state for its roots.

Sanskrit: Aswangandha. Telugu: Aswagandha, Pannerugadda.

Trade name: Aswagandha.

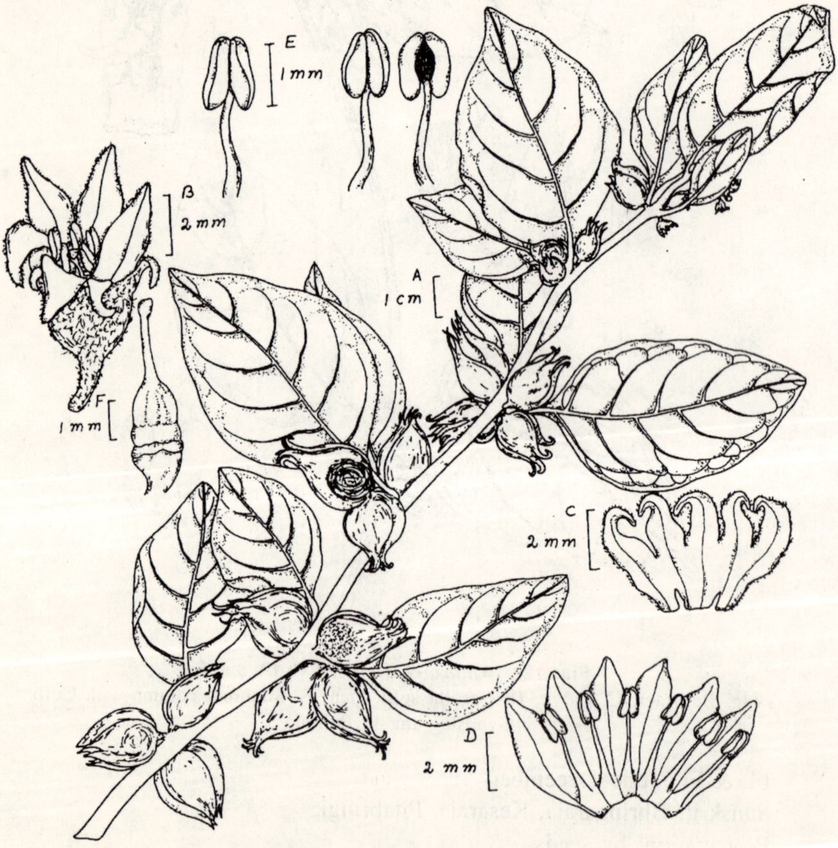

Fig. 123: *Withania somnifera* (L.) Dunal.
A. Twig; B. Flower; C. Calyx split open (Ventral view); D. Corolla split open with Stamens; E. Stamens; F. Pistil.

Medicinal uses: Roots are reported to be effective in treating leucoderma, constipation, insomnia, tissue building, nervous break down and as an aphrodisiac.

Chemistry: Roots contain alkaloids cuscohygrine, anhygrine, tropine etc.

This species grows in subtropical and tropical climate.

Cultivation: Seeds are sown or broadcast in the nursery just before the onset of rainy season. Five kgs of seeds are sufficient to provide seeds for one hectare. The seedlings are trasplanted when they are 6 weeks old, during late rainy season, in 60 cm spaced rows. Hand weeding at 30 days interval, in order to control the weeds, is necessary. Annual rainfall of 600 to 750 mm is best suited for its growth. Sandy loam or light red silts and good drainge is conducive for its growth. The plant flower and fruit during November–December. Harvesting starts in January and continues till march. The entire plant is up rooted for its roots, which are seperated from the aerial parts by cutting the stem 1–2 cm above the ground. They are then transversely cut into smaller pieces for drying. The yield of roots is 500 kg per hectare when fresh and 150 kg per hectare when dried.

Woodfordia fruticosa (L.) Kurz

Syn: *Lythrum fruticosum* L.
Woodfordia floribunda Salisb.

Family: Lythraceae.

Large woody shrubs, up to 4 m tall, bark peeling off in threads. Leaves opposite, lanceolate. Flowers greenish red, in axillary panicled cymes. Fruit oblong or ellipsoid capsule, seeds minute.

Common in all districts in deciduous forests.

Fl. & Fr.: February–May.

Sanskrit: Dhataki, Madaniyahethu. Telugu: Dhataki, Jargi, Serinji.

Trade name: Fire Flame Bush, Shiranjitha.

Propagation: by seeds.

Medicinal uses: Dried flowers are astringent and tonic. Externally, powdered flower is sprinkled over foul ulcers and wounds, for diminishing their discharge. A conflection of flower is used as stimulant and astringent, given in dysentery to check haemorrhages and leucorrhoea. Dried flowers are useful in disorders of mucous membranes, haemorrhoids and derangement of liver.

Wrightia tinctoria R.Br.

Family: Apocynaceae.

Tree, up to 15 m tall. Leaves ovate-elliptic. Flowers white, fragrant in lax terminal cymes. Fruit paired follicles, apically jointed.

Common in dry deciduous forests.

Fl. & fr.: February–August.

Sanskrit: Svetakutaja. Telugu: Ankudu, Palavareni, Tedlapala.

Trade name: Pala Indigo Plant.

Propagation: by seeds.

Medicinal uses: The bark and seeds are used in flatulence and bilious affections. A decoction of the leaves and bark is taken as a stomachic. The dried and ground bark is rubbed over body in dropsy. The seeds are said to possess aphrodisiac and anthelmintic properties.

Xanthium strumarium L.

Family: Asteraceae (Compositae).

Annual erect subshrubs, up to 1.5 m tall, stem hispid. Leaves broadly ovate, hispidulous. Flowers in terminal heads, heads monoecious, sessile, male heads many-flowered, female heads 2-flowered. Fruit oblong-ovoid achene enclosed in a 2-beaked utricle formed by the hardened involucral bracts and covered with hooked spines.

Common weed of road sides.

Fl. & Fr.: September–February.

Sanskrit: Arishta. Telugu: Marlu Mathangi.

Trade name: Cocklebur, Bur weed.

Propagation: by seeds.

Medicinal uses: It is credited with diaphoretic, emollient, diuretic and sedative properties. Its decoction is recommended in chronic malaria, leucorrhoea and urinary diseases.

Active principle: Aerial parts contain mixture of alkaloids and sesquiterpene lactones (Xanthinin, Xanthatin and Xanthumin. Xanthumin is CNS depressent.

The leaves are reported to possess astringent, alterative, antisyphlitic and diuretic properties. They are also given for scrofula and herpes. The root is a bitter tonic and has been employed in cancer and scrofula.

Ximenia americana L.

Family: Olacaceae

Armed small trees, up to 3 m tall, thorns axillary. Leaves broadly elliptic-ovate, obovate. Flowers pale green in umbellate or terminal racemes. Fruit globose drupe, orange when ripe, pulp edible, seed solitary.

Common in dry forests on stony ground.

Fl. & Fr.: March–July.

Telugu: Uranechra, Billa Nakkeru, Konda nakkeru.

Trade name: False Sandal Wood, Tallow Wood.

Propagation: by seeds.

Medicinal uses: The fruits and seeds are credited with laxative properties. The roots are said to possess cathartic properties and have been used for venereal diseases. The powdered roots are used for treating inflammation of gums and toothache. An infusion of the pulped roots is used as an anthelmintic. A decoction of the roots and leaves is used as a febrifuge and also for diarrhoea and jaundice.

Zaleya decadra (L.) Burm.f.

Syn: *Trianthema decandra* L.

Family: Aizoaceae.

Annual prostrate herb. Leaves elliptic-lanceolate or elliptic-oblong, subsucculent, papillose. Flowers pinkish in dense axillary or terminal sub-umbellate cymes or fascicles. Fruit capsule, 4-seeded, seeds orbicular or reniform, black.

Common weed in drylands and road sides.

Fl. & Fr.: June–November.

Telugu: Thellagalijeru.

Propagation: by seeds.

Medicinal uses: The root is used in hepatitis and asthma. Roots are believed to be laxative and useful in suppression of menses and in inflammation of testicles.

Zanthoxylum armatum DC

Syn: *Zanthoxylum alatum* Roxb.

Family: Rutaceae

Armed shrubs with dense foliage and pungent aromatic smell and taste. Leaves imparipinnate, petiole narrowly winged, leaves elliptic-lanceolate. Flowers in sparse panicles. Fruit pale red.

Occasional in forests in Visakhapatnam district.

Sanskrit: Tumburu, Gandhalu.

Propagation: by seeds, by cutting of half ripened shoots or root cutting.

Medicinal uses: The bark, fruits and seeds are extensively used in indigenous system of medicine as a carminative, stomachic and anthelmintic. Because of their deodorant, disinfectant and antiseptic properties, the fruits are used in dental troubles and their lotion for scabies.

Zanthoxylum limonella (Dennst.) Alston

Syn: *Zanthoxylum budrunga* Wall.
Zanthoxylum rhetsa DC.

Family: Rutaceae.

Armed, erect, deciduous trees, up to 30 m tall, prickles on the trunk and branches. Leaves crowded towards ends of branches, imparipinnate, leaflets 4–16 pairs, opposite or subopposite, ovate-lanceolate. Flowers yellow in terminal cymes. Fruit globose follicle.

Occasional in forests in north coastal Andhra pradesh.

Fl. & Fr.: June–August.

Sanskrit: Ashvaghra, Aritejani, Sutejasi. Telugu: Rhetsamanu.

Medicinal uses: The fruits are used in dyspepsia, asthma and bronchitis, hearing troubles, toothache and rheumatism. The pericarp is credited with astringent, stimulant and digestive properties. A tincture of the seeds is used in cholera.

Chemistry: The fruits yield an essential oil called Mulliam oil obtained by steam distillation.

Zingiber officinale Roscoe

Family: Zingiberaceae.

Rhizomatous perennial herb, aerial shoot leafy, up to 75 cm tall. Leaves sheathing alternate, linear lanceolate. Flowers yellow with dark purplish spots in condensed spikes. Fruit oblong capsule.

Cultivated for its rhizome.

Propagation: by rhizomes.

Sanskrit: Ardrakam. Telugu: Allamu.

Trade name: Ginger.

Propagation: by rhizomes.

Medicinal uses: Ginger is used as a stomachic, an aromatic carminative, stimulant and a flavouring agent. Ginger oil is used in mouth washes and bevarages. It is used in dyspepsia, flatulence, colds, coughs, asthma, throat complaints, gout, chronic rheumatism, headahes, cramps etc.

Active principle: Ginger contains 0.25% to 3% volatile oil possessing the aroma but not the pungency of the drug. In fact the pungency, is due to yellowish oily body, ginerol, which is odourless.

Zingber roseum (Roxb.) Roscoe

Syn: *Amomum roseum* Roxb.

Family: Zingeberaceae.

A tall, perennial, rhizomatous herb. Leaves oblong or oblong-lanceolate. Flowers red in spikes arising directly from rhizome. Fruit subglobose capsule, seeds many.

Rare in forests of north coastal districts.

Fl. & Fr.: August–September.

Ziziphus mauritiana Lam.

Syn: *Ziziphus jujuba* (L.) Gaertn.

Family: Rhamnaceae.

Much branched, small, thorny trees, up to 8 m tall, spines paired, one straight, another recurved. Leaves suborbicular, basally 3-nerved, glabrous above, rusty tomentose below. Flowers pale green in axillary pedicellate clusters. Fruit globose drupe, yellow or red when ripe, 1–2-celled.

Common in waste lands and scrub jungles.

Fl. & Fr.: October–January.

Sanskrit: Kola, Koli, Badari. Telugu: Regu.

Trade name: Indian Jujube, Chinese date.

Propagation: by seeds, air layering, bud grafting.

Medicinal uses: The seeds are edible and frequent eating of kernels, said to increase flesh and strength and cause a sedative effect, is recommended in the case of insomnia. The fruits are claimed to purify the blood and aid digestion. The roots are used in fever and to cure wounds and ulcers. The bark is used as a remedy in diarrhoea.

Active principle: A number of cyclopeptide alkaloids have been reported. Fruits are a good source of Vitamin C.

Ziziphus oenoplia (L.) Mill.

Family: Rhamanaceae.

Straggling, prickly, tomentose shrubs, prickles paired, one straight, one recurved. Leaves ovate-lanceolate, glabrous above, densely sericeous below. Flowers pale green in axillary, peduncled cymes. Fruit globose drupe, rugose, woody, dark pink at maturity edible, seed solitary, ovoid.

Common in dry forests and open bushy places.

Fl. & Fr.: October–January.

Sanskrit: Karkandauh, Srigalkoli. Telugu: Pariki.

Trade name: Jackal Jujube.

Medicinal uses: Roots used in hyper acidity and in Ascaris infection. A decoction of the bark is said to promote healing of the wounds. Fruits used for stomachache.

REFERENCES

Abdul Kareen, M. 1997. Plants in Ayurveda. FRLHT, Bangalore.

Anonymous. 2000. PDR for Herbal Medicines. Medical & Economics company, Montvale, USA.

Augustine, A.C. and L.D'Souza. 1997. Conservation of *Curculigo orchioides*. An endangered anticarcinogenic herb. In: Biotechnological Applications of Plant Tissue and Cell Culture (Eds.) G.A. Ravishankar and L.V. Venkataraman. Oxford & IBH Publishing Co. Ltd. New Delhi. pp.116–118.

Bhamare, P.B. 1998. Traditional knowledge of plants for skin ailments of Dhule and Nandurbar districts, Maharastra (India). Journ. Phytol. Res. 11: 195–196.

Bhattacharjee, S.K 2001. Hand book of medicinal plants. Pointer Publishers, Jaipur.

Bonar, A.199. Herbs-A complete guide to their cultivation and use. Tiger books International, London.

Chandel, K.P.S. 1996. Biodiversity in medicinal and aromatic plants in India (Eds.) S.Ganesh & N. Sharma. ICAR, New Delhi.

Chopra, R.N., S.L.Nayar and I.C.Chopra. 1992. Glossary of Indian Medicinal Plants. CSIR, New Delhi.

Jain, S.K.1968. Medicinal Plants. National Book Trust, New Delhi.

Jain, S.K. 1991. Dictionary of Indian Folk medicine and Ethnobotany. Deep & Deep Publications, New Delhi.

Jasrai, Y.T. and B.B.Wala. 2001. *Curculigo orchioides* Gaertn. (Kali Musli) an endangered medicinal herb. In: (Eds.) Irfan A.Khan & A.Khanum. 20001. Role of Biotechnology in Medicinal and Aromatic Plants. Ukaaz publications.

Kirtikar, K.R and B.D.Basu.1935. Indian Medicinal Plants 4.Vols.

Nair, C.K.N. and N. Mohanan. 1998. Medicinal plants of India. Nag publishers, Delhi.

Rao, P.S., K.Venkaiah and R.Padmaja. 1999. Field guide on Medicinal Plants. Andhra Pradesh Forest Department.

Shah, S. 1992. The selected Indian Medicinal Plants. Chemixil, Madras.

Sunny Oommen, D.K. Ved & R.Krishnan. 2000. Tropical Indian Medicinal Plants. Propagation methods. FRLHT, Bangalore.

COMPANIES IN MEDICINAL PLANTS TRADE

1. Abirami Botanical Corporation,
   55, P.S.S. Nadar Street, Tuticorin 628 001
   Tamil Nadu
   E-Mail: abirami@md3.vsnl.net.in
   Phone: 326810

2. Girijan Co op. Corporation Ltd.
   East Point Colony, Visakhapatnam-530017, A.P.
   Phones: 0091-891-554245, 551964
   Fax: 0091-891-554245

3. Himalaya Drug Co.
   Web site: www.thehimalayadrugco.com

4. Natco Pharma Limited, Natco House
   Road No. 2, Banjara Hills
   Hyderabad 500033, A.P.
   Phones: 274232, 272568

5. Nicholas Piramal India Limited
   In Hoechst Marion Roussel Compound
   Lal Bahadur Shastri Marg, Mulund (W)
   Mumbai-400080
   E-mail: botany@nicholaspiramal.co.in
   Phone: 022-560 6960

6. Plantex
   40-15-14, Brindavan colony, Labbipet
   Vijayawada-520010

7. Rishi Herbal Technologies
   Bangalore, Karnataka
   Phone: 080-6670797

8. Sahej Health Care Ltd.
   11-5-436/7 Red Hills Road, Lakdikapool
   Hyderabad-500004
   Phone: 0091-40-3395571
   E-mail: infoshcl@sahajhealthcare.com

9. Sanjeevani Herbal Corporation
   No. 1-116/1A Guttalabegumpet, Hyderabad-500081
   E-mail: sanjeevaniherbs@yahoo.com

10. Siris Impex, Sudarsan Bhavan
Brindavan colony, Vijayawada 520010, A.P.
E-mail: sirisvij@md3.vsnl.net.in
Phones: 471282, 473839

11. The Arya Vaidya Pharmacy (Coimbatore) Limited
326 Perumal Koil Street, Ramanathapuram
Coimbatore 641045, Tamil Nadu
E-Mail: ayurveda@vsnl.com.
Phones: 315412, 316006, 316220

12. Unicorn Natural Products
13/1 Rasulpura, Secunderabad-500003, A.P.
Phone: 91-40-814221
Fax: + 91-40-845924